SpringerBriefs in Optimization

Giorgio Fasano

Solving Non-standard Packing Problems by Global Optimization and Heuristics

 Springer

Giorgio Fasano
Thales Alenia Space
Engineering and Advanced Studies
Turin, Italy

ISSN 2190-8354   ISSN 2191-575X (electronic)
ISBN 978-3-319-05004-1 ISBN 978-3-319-05005-8 (eBook)
DOI 10.1007/978-3-319-05005-8
Springer Cham Heidelberg New York Dordrecht London

Library of Congress Control Number: 2014936305

Mathematics Subject Classification (2010): 05B40, 90C11, 90C26, 90C59, 68T20, 90C30, 90C90

Printed on acid-free paper

Springer is part of Springer Science+Business Media (www.springer.com)

Preface

This book originates from a long-lasting research effort aimed at tackling difficult non-standard packing issues arising in space engineering and logistics. In this framework the necessity of exploiting the spacecraft load capacity, as much as possible, represents a paramount challenge. This holds especially in the perspective of the manned and unmanned interplanetary missions that are going to be carried out in the near future. The experience gained in this quite peculiar context is able to suggest insights on possible extensions to several engineering and industrial sectors. They range from transportation to manufacturing, including sophisticated technological areas, such as Electronic Design Automation (EDA) and Very Large Scale Integration (VLSI).

This work is not intended to provide the readers, independently from their technical background, with a comprehensive survey on packing applications and the relevant cutting-edge methodologies. Quite a specific point of view is presented instead, reflecting the author's experience and acquired know-how. The overall *Global Optimization* (GO) approach, based on *Mixed Integer Linear/Nonlinear Programming* (MIP, MILP/MINLP) and heuristic processes, as carried out in the above mentioned context, is argued. Both the modelling, algorithmic and experimental aspects are considered, always keeping in mind possible links and synergic interactions with alternative, as well as complimentary, perspectives.

This study encloses both the author's previous consolidated work and a significant part of novel achievements, in terms of experimental outcomes, as well as model and algorithm enhancements. Drawing up this work, a systematic effort has been devoted to the harmonization and integration of both previous and current innovative material. A number of mathematical concepts are understood, as well as more specific notions (usually indicated in italics) that concern the *optimization* and *mathematical programming* context. For further in-depth clarification the reader may refer to the websites: http://mathworld.wolfram.com and http://glossary.com puting.society.informs.org.

This work most certainly focuses more on practical aspects than on theoretical rigour, referring the reader to the topical specialist literature, when necessary. It is, however, addressed both to researchers and practitioners involved in optimized packing, with strong motivation by challenging and non-conventional real-world problems.

Turin, Italy   Giorgio Fasano
January 15, 2014

About the Author

Giorgio Fasano is a researcher and practitioner at Thales Alenia Space with more than 25 years of experience in the field of optimization, in support of space engineering. He holds a M.Sc. degree in Mathematics and is a Fellow of the Institute of Mathematics and its Applications (IMA, UK). He has been awarded the Chartered Mathematician designation (IMA, UK), in addition to that of Chartered Scientist (Science Council, UK). The author is co-editor of Operations Research in Space and Air (Ciriani et al., eds., Kluwer Academic Publ., 2003), as well as of Modelling and Optimization in Space Engineering (Fasano and Pintér, eds., Springer Science + Business Media, 2013). He is the author of several specialist publications on optimization in space and was a finalist at the EURO 2007 Excellence in Practice Award. He currently works as technical manager of the CAST (Cargo Accommodation Support Tool) project, funded by the European Space Agency. His interests include Mixed Integer Programming, Global Optimization, Operations Research, Packing Optimization and Optimal Control.

Acknowledgements

The author wishes, first of all, to thank the optimization team he is in charge of at Thales Alenia Space (Turin) that made the publication of this book a feasible, albeit quite demanding, objective.

Claudia Lavopa deserves particular gratitude. She has been working, within the team, for several years, as IT specialist. During all this time, she has significantly contributed to the design and implementation of most of the software developed, suggesting very efficient ad hoc optimization strategies up to solving extremely challenging real-world instances.

Maria Chiara Vola is the team member who carried out most of the extensive test campaigns relevant to the *tetris*-like packing issue this book refers to. She offered a valuable contribution to tackling the problem of exploiting empty spaces by adding *virtual* items.

Davide Negri provided the team with a remarkable support of the design, development, as well as the test activity, both in terms of mathematical modelling and algorithm implementation. Significant results have been obtained by him, in particular as far as the heuristic aspects, relevant to the processes conceived, are concerned.

Alessandro Castellazzo has been dealing with both the linear and nonlinear approaches extensively taken account of in this book. The outcomes he has obtained are interesting, both from the development and experimental points of view. Several computational strategies thought up by him and adequately implemented in an operational framework allowed the actual solution of very large-scale exercises, definitely paving the way to further research. His editorial assistance, in addition to a huge amount of patience, was essential to the fulfilment of the whole book.

Stefano Gliozzi, senior managing consultant at IBM GBS Advanced Analytics and Optimization, offered his consolidated and extremely wide expertise in tackling large-scale MIP instances to cope with very complex tests that arose along our experimental investigation. His specialist know-how related to the advanced use of the CPLEX MIP optimizer was essential to successfully solve intricate computational cases. Also his suggestions concerning some modelling aspects resulted in being of remarkable relevance.

Janos Pintér offered very significant contributions both to the book itself and to all the topical research activity. The author is first of all grateful for his encouraging the achievement of this opus. The suggestions he provided by reviewing the whole text have certainly improved the first versions, enucleating several aspects that deserved more in-depth analysis. His specialist support in the use of the LGO nonlinear optimizer was of great aid in solving very challenging instances successfully. The recommendations he put forward, concerning in particular the formulation of the relevant nonlinear models, are expected to represent interesting directions for dedicated research in the near future.

Thanks are also due to Mario Cardano, Piero Messidoro and Annamaria Piras of Thales Alenia Space, for their sustenance of the research and development activities related to modelling and optimization in a range of space engineering applications.

The professional contribution given by Jane Alice Evans, editorial consultant, was essential to realize and harmonize the whole text. The author is extremely grateful for her invaluable commitment in reviewing the work thoroughly.

Thanks are also owed to Vaishali Daimle, previous editor in charge of this publishing project, for supporting the initiative. The joint work with Razia Amzad, present SpringerBrief editor for this book, was certainly a rewarding experience, from its initial discussions to its completion.

Turin, Italy   Giorgio Fasano
January 15, 2014

Notations

- All sets, including geometrical shapes/domains (intended as subsets of the ordinary two- to three-dimensional Euclidean space), are denoted by capital letters.
- All parameters (data) are denoted by Latin capital letters.
- All variables are designated with lower-case letters.

    Latin letters refer to continuous variables only.
    Binary (0–1) variables are indicated by Greek letters.

- The superscript '*' always refers to terms concerning balancing conditions.
- Lower bounds of variables are underlined; upper bounds are 'overlined'.
- Vectors are represented by bold letters.

Main Sets

$B = \{1, 2, 3\}$	Set of the axes of the main orthogonal reference frame
$C_i,\ i \in I$	Set of the components associated to *tetris*-like item i
D	Convex domain (polygon/polyhedron)
$E_{hi},\ i \in I,\ h \in C_i$	Set of the vertices of each component h of *tetris*-like item i
$\widehat{E}_{hi},\quad i \in I,\ h \in C_i$	E_{hi} extended by including the geometrical center of component h
I	Set of *tetris*-like items
V	Set of the vertices delimiting D
Ω	Set of all possible orthogonal rotations for each *tetris*-like item

Main Parameters

$L_{\omega\beta hi}$, $\omega \in \Omega$, $\beta \in B$, $i \in I$, $h \in C_i$ Side, parallel to the main reference frame axis w_β, of component h of *tetris*-like item i, corresponding to orientation ω

$V_{\beta\gamma}$, $\beta \in B$, $\gamma \in V$ D vertex coordinates with respect to the main reference frame

$W_{\omega\beta\eta hi}$, $\omega \in \Omega$, $\beta \in B$, $i \in I$, $h \in C_i$, $\eta \in \widehat{E}_{hi}$ Projections, on the main reference frame axes w_β, of the coordinate differences between points $\eta \in \widehat{E}_{hi}$ and the origin of the local reference frame, corresponding to orientation ω of *tetris*-like item i

Main Variables

$l_{\beta hi}$, $\beta \in B$, $i \in I$, $h \in C_i$ Projections, on the main reference frame axes w_β, of the (rectangular) parallelepiped enclosing component h of *tetris*-like item i

$w_{\beta\eta hi}$, $\beta \in B$, $\eta \in \widehat{E}_{hi}$, $i \in I$, $h \in C_i$ Coordinates, with respect to the main reference frame, of component h vertices, or its geometrical center ($\eta = 0$), relative to *tetris*-like item i

$o_{\beta i}$, $\beta \in B$, $i \in I$ Coordinates, with respect to the main reference frame, of *tetris*-like item i local reference frame origin

$\vartheta_{\omega i} \in \{0, 1\}$, $\omega \in \Omega$, $i \in I$ $\vartheta_{\omega i} = 1$ if *tetris*-like item i is loaded and has the (orthogonal) orientation $\omega \in \Omega$; $\vartheta_{\omega i} = 0$ otherwise

$\sigma_{\beta hkij}^{+/-} \in \{0, 1\}$, $\beta \in B$, $i \in I$, $h \in C_i$, $j \in I$, $k \in C_j$ $\sigma_{\beta hkij}^{+} = 1$ if the projections of component h and k of *tetris*-like items i and j respectively do not overlap on axis w_β, and k precedes h along it; $\sigma_{\beta hkij}^{-} = 1$ if the projections of component h and k of *tetris*-like items i and j respectively do not overlap on axis w_β, and h precedes k along it

$\chi_i \in \{0, 1\}$, $i \in I$ $\chi_i = 1$ if *tetris*-like item i is loaded; $\chi_i = 0$ otherwise

Contents

Chapter 1
Non-standard Packing Problems: A Modelling-Based Approach

The general subject of packing objects, exploiting the available volume, as much as possible, has represented, for centuries, or even longer, an extremely tricky task. This issue seems trivial, until one encounters it. The question arose, for instance, when dealing with cannon ball stowage in ancient vessels. It is not surprising at all that it gained the role of the packing issue par excellence, when Hilbert announced his eighteenth problem (to date resolved by computer-assisted proof, e.g. Gray 2000). It concerned the accommodation of equal spheres, attaining the maximum density.

Paramount effort has been carried out, and is ongoing, to dominate extremely challenging overall packing problems, from the theoretical point of view. Well-known directions of speculative investigations include infinite dimensional space issues. For instance, we could consider the packing of Platonic solids in the ordinary Euclidean space and of spheres in higher dimensions. Further examples concern finite space questions, such as those of placing squares/circles or cubes/spheres into regular figures (e.g. http://mathworld.wolfram.com).

An unquestionably much more practical slant is instead underlined in the operations research and computational geometry frameworks. In such a context, the role of the numerical approach to look into high-quality (albeit usually nonproven optimal) solutions to even more complex, real-world packing problems is emphasized. This is most definitely the point of view of this work.

There is vast specialist literature on multidimensional packing from a numerical optimization standpoint. It is, therefore, not intended to be surveyed here. The reader may refer to some comprehensive overviews (e.g. Cagan et al. 2002; Dyckhoff et al. 1997; Ibaraki et al. 2008). As is known, a significant part of the topical bibliography focuses on the orthogonal placement of rectangles/parallelepipeds into rectangles/parallelepipeds (e.g. Faroe et al. 2003; Fekete and Schepers 2004; Fekete et al. 2007; Martello et al. 2000; Pisinger 2002), even if several works also consider different typologies of packing issues (e.g. Addis et al. 2008a; Birgin et al. 2006; Egeblad et al. 2007; Gomes and Olivera 2002; Scheithauer et al. 2005; Teng et al. 2001).

G. Fasano, *Solving Non-standard Packing Problems by Global Optimization and Heuristics*, SpringerBriefs in Optimization, DOI 10.1007/978-3-319-05005-8_1, © Giorgio Fasano 2014

Several versions of two-/three-dimensional packing problems can be differently classified, depending on the specific optimization criterion adopted. When, for instance, the number of containers is fixed and the total load has to be maximized (e.g. in terms of its volume/value), the relevant model is referred to as a *knapsack* problem (e.g. Caprara and Monaci 2004; Egeblad and Pisinger 2006, 2009; Fekete and Schepers 1997). It is reduced to the *single container* one, when only one container is available (e.g. Bortfeldt et al. 2012; Kang et al. 2010; Parreño et al. 2008).

The issue of loading a set of given objects, whilst minimizing the number of containers to utilize (or, more in general, their total volume/cost), is referred to as the bin packing problem (e.g. Lodi et al. 2010; Martello et al. 2000; Pisinger and Sigurd 2007).

Further questions concern the 'reduction' of the container. This is the case, in particular, of the *strip packing* problem (e.g. Iori et al. 2003; Kenmochi et al. 2009; Zhang et al. 2006), where a single dimension of the (rectangle/parallelepiped-shaped) domain has to be minimized. Another class of interesting issues consists of the (rectangle/parallelepiped-shaped) domain (area/volume) minimization problem (e.g. Li et al. 2002; Pan and Liu 2006). This is of importance in several applications, ranging from manufacturing and logistics to electronic design (e.g. *floor-planning* in very large scale integration, VLSI). Still related to the container volume minimization, it is worth mentioning the issue of the *sphere packing* in optimized spheres (e.g. Kampas and Pintér forthcoming).

Remarkable effort has been dedicated to tackling several kinds of packing problems algorithmically, often by adopting general *meta-heuristics* or dedicated heuristics (e.g. Allen et al. 2011; Bennell et al. 2013; Bennel et al. 2013; Bennell and Oliveira 2009; Bortfeldt and Gehring 2001; Bortfeldt et al. 2003; Burke et al. 2006, 2010; Coffman et al. 1997; Dowsland et al. 2006; Gehring and Bortfeldt 2002; Gomes and Olivera 2002; Gonçalves and Resende 2012; Hopper and Turton 2001, 2002; López-Camacho et al. 2013; Mack et al. 2004; Oliveira et al. 2000; Pisinger 2002; Ramakrishnan et al. 2013; Terashima-Marín et al. 2010; Wang et al. 2008; Yeung and Tang 2005). Nonetheless, modelling-based approaches have also been investigated (e.g. Allen et al. 2012; Chen et al. 1995; Chernov et al. 2010; Fasano 1989; Fischetti and Luzzi 2009; Hadjiconstantinou and Christofides 1995; Kallrath 2009; Padberg 1999; Pisinger and Sigurd 2005). These works refer to the overall context of *Mathematical Programming*, including *mixed-integer (linear) programming* (MILP, MIP) and *mixed-integer nonlinear programming* (MINLP).

The underlying theme examined here originates from a long-lasting research work devoted to tackling complex non-standard packing issues arising in space applications. These usually concern both design and logistics aspects. In this sector, the necessity of exploiting the spacecraft load capacity, as much as possible, presents the engineers with a paramount challenge. This is foreseen especially in the perspective of the extremely demanding missions that are going to be carried out in the near future.

Generally, the volume or the mass of the loaded cargo has to be maximized. Other optimization criteria, however, can also be stated, depending on the specific mission scenarios to deal with. In any case, very tough accommodation rules have to be taken into account, in compliance with demanding requirements relative to safety, ergonomic and operational concerns.

Tight balancing conditions, deriving from control specifications, are usually posed at an overall system level (i.e. the whole spacecraft). However, the requirement of considering them also when loading each single internal container (such as, for instance, racks or bags) is quite often needed. The space-system internal geometries are normally very intricate. As a consequence, in order to exploit each available volume, as much as possible, the shape of the adopted containers themselves can become quite peculiar. This occurs, for instance, when dealing with the cargo accommodation of the European Automated Transfer Vehicle (ATV, ESA, cf. http://www.esa.int). In such a case, some specific bags have curved surfaces to fit with the shape of the racks they have to be accommodated into that is, itself, determined by the cylindrical form of the carrier.

A specific class of 'hard' non-standard packing problems with additional conditions arises. All this is determined by balancing conditions, the shapes of domains and objects, the possible presence of internal *separation* planes, or *structural* elements, not to mention complicated accommodation rules. This situation can arise in space engineering and logistics. Despite the specificity of the context, however, it is, in more or less similar versions, susceptible to several real-world applications. This happens also in very different frameworks.

Balancing conditions and complex geometries, for instance, are increasingly important subjects in the high-speed transportation system sector. Complex engineering structures (e.g. oil rigs), even if related to quite different operational scenarios, pose similar problems. Non-standard packing issues have to be considered daily in manufacturing, even if not necessarily in the presence of balancing conditions. This occurs, often, just in a two-dimensional context (e.g. Electronic Design Automation, EDA and VLSI). This work discusses in depth some real-world packing scenarios, from an application perspective. It is aimed at introducing an efficient methodology to solve non-trivial problems in practice.

In several applicative contexts, items can often be assumed to be (rectangular) parallelepipeds, without significant loss of information. Nevertheless, generally, such an approximation does not hold, especially when dealing with large and complex items. Similar considerations can, moreover, be made, considering the container shape, since frequently it is not just a (rectangular) parallelepiped. The presence of additional geometric and operational conditions presents further challenges.

Remarkable works, concerning non-standard packing problems, are available. In the author's opinion, however, this topic definitely deserves much more commitment, also in consideration of the increasing demand generated by the real-world context. This is the essential motivation inspiring the drawing up of the present work.

When dealing with non-standard packing problems, with overall conditions, such as balancing, the simplistic approach (adopted by several packing algorithms) of placing items one at a time is scarcely efficient. A strongly nonlocal viewpoint is therefore highly desirable, also in consideration of the outstanding results recently achieved in the framework of *global optimization* (GO, consult, e.g. Addis et al. 2008b; Castillo et al. 2008; Floudas et al. 2005; Floudas and Pardalos 1990, 2001; Floudas et al. 1999; Horst and Pardalos 1995, 1997; Horst and Tuy 1996; Kallrath 1999, 2008; Liberti and Maculan 2005; Locatelli and Raber 2002; Pardalos and Resende 2002; Pardalos and Romeijn 2002; Pintér 1996, 2006, 2009; Rebennack et al. 2009).

A modelling-based philosophy, as opposed to a pure algorithmic one, has been looked into, characterizing the whole approach followed hereinafter. GO represents therefore a first highlight. The packing problems in general, moreover, even when posed in quite an elementary version (e.g. the placement of simple boxes in a container box, without any additional conditions) are well known for being *NP-hard*. As a consequence, no deterministic methodology to successfully solve the problem to optimality is expected. An overall heuristic point of view is therefore a second key characteristic of this volume. In particular, a joint use of GO, based on MIP/MINLP formulations, and heuristic procedures is emphasized.

The concept of tetris-like item is introduced, representing a fundamental reference paradigm for the generic approach proposed here. It generalizes the original idea of tetris item, deriving, in turn, from that of *polyomino* (see Golomb 1994). Either three- or two-dimensional tetris-like items are considered, and, differently from the original concept, they are not supposed to have integer side lengths.

This notion is, by itself, quite interesting, as it is adequate to represent a wide range of real-world objects, in a streamlined but sufficiently realistic way. In several applications, indeed, one has to deal with quite complex objects, characterized by intricate shapes. Considering items, as a whole, just in terms of their smallest enclosing boxes would clearly result, in most cases, extremely restrictive. In any accommodation practical problem, where quite an efficient exploitation of the overall volume available is a mandatory task, this would be of no use at all. Substituting complex objects with tetris-like items, i.e. clusters of mutually perpendicular cuboids (rectangular parallelepipeds), could make their representation much more realistic, even if simplified and thus still approximate.

The original object is therefore partitioned into parts, enclosing each in a cuboid. Obviously, the bigger (i.e. refined) the partition is, the more realistic the representation results. Figure 1.1 shows a (not too sophisticated) tetris-like approximation of a real-world object. Similarly to the case concerning the items to load, structural elements, equipment/devices and clearance/accessibility regions, inside the container, may well benefit from this representation, as illustrated by Fig. 1.2, referring to the internal part of a space module. Reinforcements of the cylindrical structure are present, together with some electronic devices that are supposed to be protected by forbidden zones.

In addition to what mentioned above, the relevant modelling features of the tetris-like representation are quite suitable for MIP formulations that provide a

 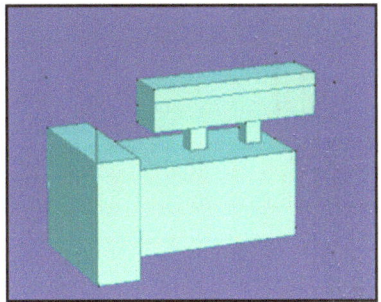

Fig. 1.1 Representation of complex objects with tetris-like items

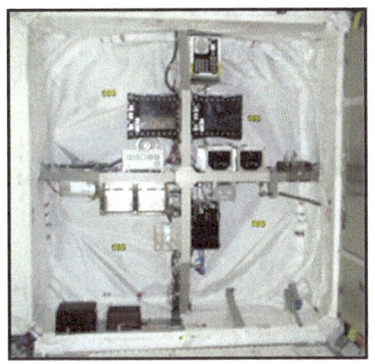 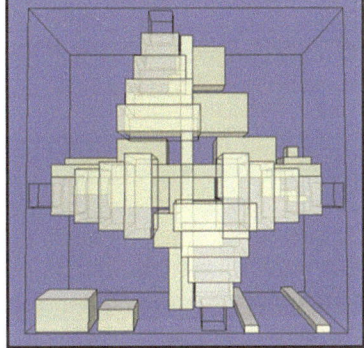

Fig. 1.2 Internal devices and forbidden zones approximated by tetris-like items

linear-based overall structure, obviously beneficial to a GO approach. Moreover, its MIP formulation is able to cover a non-negligible number of additional conditions.

This volume both reviews the author's previous works (e.g. Fasano 2008, 2013) and introduces new research outcomes, establishing a basis for further investigation and development.

The second chapter discusses, at quite a detailed level, a general mathematical model for the orthogonal packing of three-dimensional tetris-like items within a convex domain (polyhedron). Some critical aspects are pointed out, suggesting how it is quite easy to overcome them. A number of additional conditions are looked into, including the prefixed position/orientation of subsets of items, the presence of 'holes' or forbidden zones as well as of *separation* planes and *structural* elements, relative distance bounds and *static/dynamic* balancing requirements.

The corresponding *feasibility* subproblem is discussed in the third chapter. It consists of the special case taking place when no optimization criterion (e.g. the total volume maximization) is selected a priori, and all items have to be loaded. This situation can be profitably exploited by introducing an ad hoc *objective* function, aimed at facilitating the resolving process in finding *integer-feasible* solutions. Both linear and nonlinear readjustments of the general MIP model are considered. The third chapter also outlines the issue of *tightening* the general MIP model, by introducing implications and *valid* inequalities, suitable, in particular, for a dedicated *branch-and-cut* approach.

As the general MIP model is extremely tough to solve, even when not too large-scale instances are involved, an MIP-based heuristic point of view is described in the fourth chapter. There, the basic concept of *abstract configuration* is enucleated. It essentially consists of a set of item-item relative positions, feasible in any unbounded domain. The *feasibility* sub-models are profitably adopted to generate 'good' *abstract configurations*. The heuristic approaches delineated in this chapter are founded on recursive generations of these.

The fifth chapter is devoted to the experimental results, obtained to date, relevant to a real-world application framework. The sixth explores both extensions of the general MIP model and nonlinear (MINLP) formulations, in order to tackle two further non-standard packing issues. The first concerns the creation of possible *virtual* items, to exploit the empty spaces of a container, already partially loaded with tetris-like items. This aspect is of importance in several applications. The second issue deals with the non-orthogonal placement of polygons with (continuous) rotations in a convex domain (polygon). Also in this case, a GO-based heuristic approach is proposed. It is aimed at finding a 'good' approximate solution susceptible to further local refinement by more sophisticated formulations, such as the one based on the Stoyan's Φ-functions (e.g. Stoyan et al. 2004). The tetris-like item model is advantageously exploited to provide the MINLP solution process with a 'good' starting solution.

The last chapter concludes the volume providing some insights on prospective enhancements, in terms of further experimental analysis needed, but also from the modelling and development point of view, including extended applications (one in particular dealing with scheduling problems).

Chapter 2
Tetris-like Items

The packing of tetris-like items, i.e. clusters of mutually orthogonal rectangular parallelepipeds, inside a given domain, is discussed here; see Fig. 2.1. Orthogonal rotations are admitted and additional conditions can be present. Before introducing the general problem and its mathematical formulation, the following definition is provided as a fundamental concept.

Definition 2.1 A tetris-like item is a set of rectangular parallelepipeds positioned orthogonally, with respect to an (orthogonal) reference frame. This is called 'local' and each parallelepiped 'component'.

In the following, 'tetris-like item' will usually be simply denoted as 'item', if no ambiguity occurs. Similarly, 'rectangular parallelepipeds' are indicated as 'parallelepipeds'.

The term 'domain' refers to a subset of the *three-dimensional Euclidean space $\boldsymbol{R}^3$*. Convex domains are mainly considered, providing the proper specifications explicitly, when otherwise. The general problem is examined first (Sect. 2.1), discussing some possible criticalities (Sect. 2.2), before investigating the issue of modelling a set of additional conditions (Sect. 2.3).

2.1 General Problem Statement and Mathematical Model Formulation

This section looks upon a first basic statement of the tetris-like packing issue, as an extension of the classical single *container loading* problem. The issue of placing small boxes into a big one has consolidated mathematical models. The formulation usually referred to as *space-indexed* is based on the container discretization (e.g. Beasley 1985; Hadjiconstantinou and Christofides 1995). The relative MIP model provides very strong bounds (see Allen et al. 2012), as it also occurs for similar discretized formulations for scheduling problems (corresponding to

G. Fasano, *Solving Non-standard Packing Problems by Global Optimization and Heuristics*, SpringerBriefs in Optimization, DOI 10.1007/978-3-319-05005-8_2, © Giorgio Fasano 2014

Fig. 2.1 Tetris-like item
packing into a convex
domain (polyhedron)

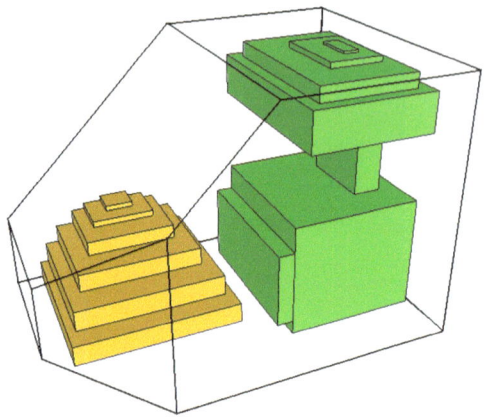

one-dimensional packing, e.g. Pan and Shi 2007). This characteristic, in several
cases, greatly makes up for the discretization. This holds, in particular, when all the
items involved have integer side lengths (e.g. unit squares/cubes). The extension of
this *space-indexed* formulation to the accommodation of the tetris-like items issue
discussed in this chapter would be quite straightforward.

A *non-space-indexed* paradigm is considered in this work. A corresponding
mathematical model, expressed in terms of *mixed-integer linear programming*
(MILP), is formulated (it is usually denoted as the general MIP model, when no
ambiguity occurs, omitting the specification 'linear').

To state the problem, we shall consider a set of N items, each identified by an
associated local reference frame. This set is denoted by I. A (bounded) convex
domain D, consisting of a polyhedron (see Fig. 2.1), is considered. It is associated to
a given orthogonal reference frame, denoted in the following as main. The problem
is that of placing items into D, maximizing the loaded volume (or mass), with the
following positioning rules:

- *Each local reference frame axis has to be positioned orthogonally, with respect
  to the main frame* (*orthogonality* conditions).
- *For each item, each component has to be contained within D* (*domain*
  conditions).
- *Components of different items cannot overlap* (*non-intersection* conditions).

This problem can easily be formulated as an MIP (Fasano 2008). When dealing
with tetris-like items, each one consisting of a single *component* only and a domain
consisting of a parallelepiped, the tetris-like item general problem stated above
reduces to the classical *container loading* issue (e.g. Bortfeldt and Wäscher 2012).
Its MIP formulation can be found, with possible variations, in some previous works
(e.g. Chen et al. 1995; Fasano 1989, 1999, 2003, 2004; Padberg 1999; Pisinger and
Sigurd 2005).

To formulate the general MIP model in question, the main orthogonal reference
frame, with origin O and axes w_β, $\beta = \{1, 2, 3\} = B$, is defined. Each local
reference frame, associated to every item i, is chosen, without loss of generality,

Fig. 2.2 Tetris-like item rotations around a single axis

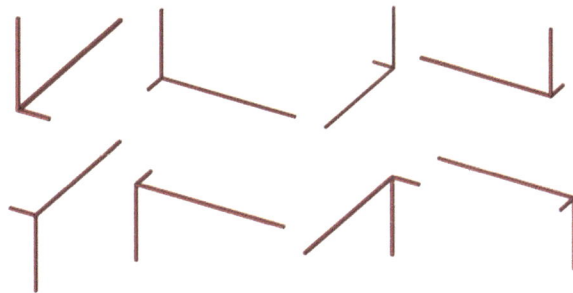

so that all item *components* lie within its first octant. Its origin coordinates, with respect to the main reference frame, are denoted in the following by $o_{\beta i}$. We shall then introduce the set Ω of all possible orthogonal rotations, admissible for any local reference frame, with respect to the main one. It is easily seen that they are 24 in all, since items are, in general, asymmetric objects.

This is illustrated by Fig. 2.2, where an item, consisting of three mutually orthogonal components, is considered. The components have lengths of 1, 3 and 9 units respectively. The component of length 3 units is parallel to the vertical axis of the observer reference frame. Two sub-cases are considered: in one (corresponding to the four images above) the item is *up-oriented*, whilst in the other (corresponding to the four images below) it is *down-oriented*. As can be seen from the figure, four orthogonal (clockwise) rotations (around the vertical axis) are associated to each sub-case, so that when the component of length 3 units is vertical (either *up-oriented* or *down-oriented*), eight relative rotations have to be taken into account. The same holds when either the component of length 1 unit or the one of length 9 units assumes the vertical position, so that the total number of orthogonal orientation is 24.

In the following, the set of *components* associated to the generic item i is denoted by C_i. We shall introduce, for each item i, the set E_{hi} of all (eight) vertices associated to each of its *component h*. An extension of this set is obtained by adding to E_{hi} the geometrical centre of *component h*. This extended set is denoted in the following by $\widehat{E}_{hi}$. For each item i and each possible orthogonal orientation $\omega \in \Omega$, we define the following binary (0–1) variables:

$\chi_i \in \{0, 1\}$, with $\chi_i = 1$ if item i is picked, $\chi_i = 0$ otherwise;

$\vartheta_{\omega i} \in \{0, 1\}$, with $\vartheta_{\omega i} = 1$ if item i is picked and has the orthogonal orientation $\omega \in \Omega$, $\vartheta_{\omega i} = 0$ otherwise.

The above *orthogonality* conditions can be expressed as follows:

$$\forall i \in I \quad \sum_{\omega \in \Omega} \vartheta_{\omega i} = \chi_i, \tag{2.1}$$

$$\forall \beta \in B, \forall i \in I, \forall h \in C_i, \forall \eta \in \widehat{E}_{hi}$$
$$w_{\beta \eta hi} = o_{\beta i} + \sum_{\omega \in \Omega} W_{\omega \beta \eta hi} \vartheta_{\omega i}. \tag{2.2}$$

Here $w_{\beta\eta hi}$ ($\forall \eta \in \widehat{E}_{hi}$) are the vertex coordinates, with respect to the main reference frame, of *component h*, or its geometrical centre ($\eta = 0$), relative to item i; $W_{\omega\beta\eta hi}$ are the projections on the axes w_β of the coordinate differences between points $\eta \in \widehat{E}_{hi}$ and the origin of the local reference frame, corresponding to orientation ω of item i.

The *domain* conditions are expressed as follows:

$$\forall \beta \in B, \forall i \in I, \forall h \in C_i, \forall \eta \in E_{hi}$$
$$w_{\beta\eta hi} = \sum_{\gamma \in V} V_{\beta\gamma} \lambda_{\gamma\eta hi}, \tag{2.3}$$

$$\forall i \in I, \forall h \in C_i, \forall \eta \in E_{hi} \quad \sum_{\gamma \in V} \lambda_{\gamma\eta hi} = \chi_i. \tag{2.4}$$

Here $w_{\beta\eta hi}$ ($\eta \in E_{hi}$) are the vertex coordinates, with respect to the main reference frame, of *component h* relative to item i; V is the set of vertices delimiting D and $V_{\beta\gamma}$ are their coordinates (assumed as non-negative, with no loss of generality) and $\lambda_{\gamma\eta hi}$ are non-negative variables. These conditions correspond to the well-known necessary and sufficient conditions for which a point belongs to a convex domain.

The *non-intersection* conditions are represented by the constraints below:

$$\forall \beta \in B, \forall i, j \in I / i < j, \forall h \in C_i, \forall k \in C_j$$
$$w_{\beta 0 hi} - w_{\beta 0 kj} \geq \frac{1}{2} \sum_{\omega \in \Omega} \left(L_{\omega\beta hi} \vartheta_{\omega i} + L_{\omega\beta kj} \vartheta_{\omega j} \right) - D_\beta \left(1 - \sigma^+_{\beta hkij} \right), \tag{2.5a}$$

$$\forall \beta \in B, \forall i, j \in I / i < j, \forall h \in C_i, \forall k \in C_j$$
$$w_{\beta 0 kj} - w_{\beta 0 hi} \geq \frac{1}{2} \sum_{\omega \in \Omega} \left(L_{\omega\beta hi} \vartheta_{\omega i} + L_{\omega\beta kj} \vartheta_{\omega j} \right) - D_\beta \left(1 - \sigma^-_{\beta hkij} \right), \tag{2.5b}$$

$$\forall i, j \in I / i < j, \forall h \in C_i, \forall k \in C_j$$
$$\sum_{\beta \in B} \left(\sigma^+_{\beta hkij} + \sigma^-_{\beta hkij} \right) \geq \chi_i + \chi_j - 1, \tag{2.6}$$

where D_β are the sides (parallel to the main reference frame) of the parallelepiped, of minimum dimensions, containing D (*minimum enclosing parallelepiped*); $w_{\beta 0 hi}$ and $w_{\beta 0 kj}$ are the centre coordinates, with respect to the main reference frame, of *components h* and k of items i and j respectively; $L_{\omega\beta hi}$ and $L_{\omega\beta kj}$ are their side projections on axes w_β, corresponding to orientation ω; $\sigma^+_{\beta hkij}$ and $\sigma^-_{\beta hkij} \in \{0, 1\}$. Inequalities (2.5a) and (2.5b) state that if, for any pair of *components h* (of i) and k (of j), a variable σ is equal to one, then the corresponding *non-intersection* constraint is made active; otherwise it becomes redundant. The condition $\sigma^+_{\beta hkij} = 1$ means that k precedes h, with respect to the axis w_β and vice versa if $\sigma^-_{\beta hkij} = 1$. When both i and j

are picked, the relative inequality (2.6) guarantees that at least one σ is equal to one, and this means that, at least, one *non-intersection* constraint holds.

The *objective* function has the following expression:

$$\max \sum_{i \in I} K_i \chi_i, \tag{2.7}$$

where K_i is either the volume V_i or the mass M_i of item i. The total volume loaded is denoted by v, with $v = \sum_{i \in I} V_i \chi_i$, whilst analogously, m, with $m = \sum_{i \in I} M_i \chi_i$, refers to the overall mass.

Remark 2.1 The constants D_β can be interpreted as the classical big-*Ms* of the mathematical programming literature and the inequalities (2.5a) and (2.5b) as big-*M* constraints. It is indeed quite easy to rearrange each of them in the more usual general form $\sum_l K_l u_l \leq K_0 + (1 - \varsigma)K$, where K_l, K_0 are constants, u_l are continuous variables, $\varsigma \in \{0, 1\}$ and K is the corresponding big-*M*. Each big-*M* could be theoretically substituted with any constant, sufficiently big to make the relative constraint redundant when the corresponding ς is zero. In the whole MIP/MINLP context, it is, however, well known that there is computational advantage in making each big-*M* as small as possible, without excluding any integer-feasible solutions (e.g. Williams 1993).

A property of interest, dealing with the issue of *tightening* the big-*Ms* appearing in (2.5a) and (2.5b), is briefly looked upon here below. Prior to introducing Proposition 2.1, we shall define the concept of *external component*, i.e. a *component* with, at least, one side adjacent to the *minimum enclosing parallelepiped*, enveloping the corresponding tetris-like item.

Proposition 2.1 *Given that the domain D is a parallelepiped, for any pair of external components h and k, belonging to two different tetris-like items, the (big-M) terms D_β, appearing in (2.5a) and (2.5b), cannot be further tightened.*

Proof Consider any two items i, j and any relative external components h and k respectively. We shall write constraints (2.5a) and (2.5b) in the form

$$\forall \beta \in B, \forall i, j \in I / i < j, \forall h \in C_i, \forall k \in C_j$$

$$w_{\beta 0hi} - w_{\beta 0kj} \geq \frac{1}{2} \sum_{\omega \in \Omega} \left(L_{\omega \beta hi} \vartheta_{\omega i} + L_{\omega \beta kj} \vartheta_{\omega j} \right) - K_{\beta hkij}^+ \left(1 - \sigma_{\beta hkij}^+ \right),$$

$$\forall \beta \in B, \forall i, j \in I / i < j, \forall h \in C_i, \forall k \in C_j$$

$$w_{\beta 0kj} - w_{\beta 0hi} \geq \frac{1}{2} \sum_{\omega \in \Omega} \left(L_{\omega \beta hi} \vartheta_{\omega i} + L_{\omega \beta kj} \vartheta_{\omega j} \right) - K_{\beta hkij}^- \left(1 - \sigma_{\beta hkij}^- \right),$$

where $K^+_{\beta hkij}$ and $K^-_{\beta hkij}$ are positive constants. If $\sigma^+_{\beta hkij} = 1$, the non-intersection constraint between h and k, with respect to the corresponding axis w_β becomes active, as the multiplier of $K^+_{\beta hkij}$ reduces to zero. If, instead, $\sigma^+_{\beta hkij} = 0$, the inequality below must hold for any possible position of i and j in D:

$$K^+_{\beta hkij} \geq -w_{\beta 0hi} + w_{\beta 0kj} + \frac{1}{2} \sum_{\omega \in \Omega} \left(L_{\omega\beta hi}\vartheta_{\omega i} + L_{\omega\beta kj}\vartheta_{\omega j} \right).$$

To prove the proposition, it is therefore sufficient to consider the extreme case, where item j *minimum enclosing parallelepiped*, is, with respect to the corresponding side D_β, at its upper bound and item i *minimum enclosing parallelepiped* at the lower one. Denoting, for any rotation ω and ω' of i and j, respectively, by $\bar{L}_{\omega\beta i}$ and $\bar{L}_{\omega'\beta j}$, the side projections of their *minimum enclosing parallelepipeds* on w_β, the following inequality must hold:

$$K^+_{\beta hkij} \geq D_\beta - \bar{L}_{\omega'\beta j} + W_{\omega'\beta 0kj} + \frac{1}{2}L_{\omega'\beta kj} - \bar{L}_{\omega\beta i} + W_{\omega\beta 0hi} + \frac{1}{2}L_{\omega\beta hi}.$$

As this is requested for any ω and ω', the condition is true, in particular, also when $-\bar{L}_{\omega'\beta j} + W_{\omega'\beta 0kj} + \frac{1}{2}L_{\omega'\beta kj} - \bar{L}_{\omega\beta i} + W_{\omega\beta 0hi} + \frac{1}{2}L_{\omega\beta hi} = 0$, occurring when both h and k attain the contact condition (with respect the domain sides perpendicular to D_β). The same reasoning occurs, obviously, for $\sigma^-_{\beta hkij}$, so that neither $K^+_{\beta hkij}$ nor $K^-_{\beta hkij}$ can be smaller than D_β.   □

Remark 2.2 The proof of Proposition 2.1 suggests how to determine, albeit in a much more complicated way, the smallest big-*M*s when the domain D is not just a parallelepiped but a more general polyhedron. For this purpose, let us introduce the terms $\underline{W}_{\omega\beta 0hi}$, representing the minimum value that $w_{\beta 0hi}$ can attain (in D) when item i has the orientation ω, and $\overline{W}_{\omega\beta 0hi}$, representing the maximum value that $w_{\beta 0hi}$ can attain (in D) when item i has the orientation ω.

The   minimum $\underline{K}^+_{\beta hkij}$ is   hence   determined   by   the   expression $\underline{K}^+_{\beta hkij} = \max_{\omega, \omega' \in \Omega} \left\{ -\underline{W}_{\omega\beta 0hi} + \overline{W}_{\omega'\beta 0kj} + \frac{1}{2}\left(L_{\omega\beta hi} + L_{\omega'\beta kj} \right) \right\}$ and analogously for the corresponding minimum $\underline{K}^-_{\beta hkij}$. The above considerations hold both for *external* and non-*external components*, including, for these, the case when D is a parallelepiped.

Special case of single parallelepipeds (single-component items)
The special case concerning single-*component* items is introduced here. The resulting single parallelepipeds are assumed to be of homogeneous density, so that the problem is greatly simplified, as six orthogonal rotations, with respect to the main reference frame, are sufficient to determine their actual orientation. A further simplification is carried out, restricting the domain D to be a parallelepiped.

Posing $\alpha \in \{1, 2, 3\} = A$, for each parallelepiped i, denote by $L_{\alpha i}$ its sides, supposing, with no loss of generality, $L_{1i} \leq L_{2i} \leq L_{3i}$ (as this assumption can be of use in extended versions of this model). The coordinates of its centre, with respect to the main reference frame, are indicated as $w_{\beta i}$. The domain D is a parallelepiped with sides D_β parallel to the main reference frame axes w_β, respectively. A vertex of D is, moreover, supposed to be coincident with its origin O and D lies within the first octant. For each item i, the binary variables $\delta_{\alpha\beta i} \in \{0, 1\}$ are introduced, with the meaning $\delta_{\alpha\beta i} = 1$ if $L_{\alpha i}$ is parallel to the axis w_β and $\delta_{\alpha\beta i} = 0$ otherwise.

The general *objective* function (2.7) is kept unchanged, whilst the overall conditions are rewritten as follows:

Orthogonality constraints:

$$\forall \alpha \in A, \forall i \in I \quad \sum_{\beta \in B} \delta_{\alpha\beta i} = \chi_i, \tag{2.8a}$$

$$\forall \beta \in B, \forall i \in I \quad \sum_{\alpha \in A} \delta_{\alpha\beta i} = \chi_i. \tag{2.8b}$$

Domain constraints:

$$\forall \beta \in B, \forall i \in I$$
$$0 \leq w_{\beta i} - \frac{1}{2} \sum_{\alpha \in A} L_{\alpha i} \delta_{\alpha\beta i} \leq w_{\beta i} + \frac{1}{2} \sum_{\alpha \in A} L_{\alpha i} \delta_{\alpha\beta i} \leq D_\beta \chi_i. \tag{2.9}$$

Non-intersection constraints:

$$\forall \beta \in B, \forall i, j \in I / i < j$$
$$w_{\beta i} - w_{\beta j} \geq \frac{1}{2} \sum_{\alpha \in A} \left(L_{\alpha i} \delta_{\alpha\beta i} + L_{\alpha j} \delta_{\alpha\beta j} \right) - \left(1 - \sigma_{\beta ij}^+ \right) D_\beta, \tag{2.10a}$$

$$\forall \beta \in B, \forall i, j \in I / i < j$$
$$w_{\beta j} - w_{\beta i} \geq \frac{1}{2} \sum_{\alpha \in A} \left(L_{\alpha i} \delta_{\alpha\beta i} + L_{\alpha j} \delta_{\alpha\beta j} \right) - \left(1 - \sigma_{\beta ij}^- \right) D_\beta, \tag{2.10b}$$

$$\forall i, j \in I / i < j$$
$$\sum_{\beta \in B} \left(\sigma_{\beta ij}^+ + \sigma_{\beta ij}^- \right) \geq \chi_i + \chi_j - 1, \tag{2.11}$$

where $\sigma_{\beta ij}^+$ and $\sigma_{\beta ij}^- \in \{0, 1\}$.

2.2   Excluding Non-realistic Solutions

The general MIP model discussed in Sect. 2.1 can be subject to criticism, whenever some practical aspects cannot be neglected. First of all, it does not include constraints that prevent items to assume a 'floating' position within the container. This flaw is scarcely influential in some contexts, such as that of bag loading in space logistics, where the internal empty spaces are usually filled with packaging material (e.g. bubble wrap or foam). Nonetheless, it can be non-negligible in several different applications.

For the sake of simplicity, referring to the case of single parallelepipeds, a 'trick' to overcome this difficulty in practice can easily be adopted. To this purpose, it is sufficient to carry out a two-step optimization process, the first aimed at finding what the maximum cargo (in terms of volume or mass) is and the second at lowering the position of the items towards the container floor, as much as possible.

Let us suppose that the general MIP model is aimed at maximizing the loaded volume (similar considerations would hold in the case of the mass) and denote by $\hat{v}$ the (optimal) value, obtained at the first step. The second is executed by adopting a variation of the general MIP. The constraint $\sum_{i \in I} V_i \chi_i = \hat{v}$ is therefore added, imposing that the total volume must be equal to that obtained (as an optimal solution) in the first step. The *objective* function $\min \sum_{i \in I} w_{3i}$, replaces (2.7) (axes w_3 is assumed as the vertical one). This will lower the position of each item, reducing, as a consequence, the undesired 'floating' position effect (if an optimal solution, in particular, is found, no item can stay totally suspended any longer).

Remark 2.3 If, in the first step, only a suboptimal (or a nonproven optimal) solution is found, the inequality $\sum_{i \in I} V_i \chi_i \geq \hat{v}$ can substitute the corresponding equation.

As an alternative to what is outlined above, a more sophisticated approach could be followed. This may be done by providing the main general MIP model with further constraints, to make the packing solutions as realistic as possible, taking into account both layer and stability conditions (see Sect. 2.3.5).

In addition to the criticality due to the aforementioned 'floating' positions, a non-trivial issue may arise when in the presence of 'closed' tetris-like items, as shown in Fig. 2.3. In some cases, to exclude solutions that are not feasible from the practical point of view, particular precautions should be taken. This may be done by stating special constraints not contemplated by the general MIP model of Sect. 2.1.

To explain the concept, let us consider a very simple example (see Fig. 2.3), by introducing a pair of (identical) items i and j, each composed of four parallelepipeds (*components*) forming a 'squared' ring (homeomorphic to the classical topological figure of the torus).

As is immediately gathered, the *non-intersection* conditions (2.5a), (2.5b) and (2.6), reported in Sect. 2.1, are not sufficient to avoid situations such as the one illustrated in Fig. 2.3. They, indeed, prevent the intersection of each *component*

Fig. 2.3 Concatenated
'squared' rings

Fig. 2.4 'Squared' ring
(*left*) with additional
(zero-mass internal)
component (*right*)

Fig. 2.5 Acceptable
solution

of i with every one of j, and this holds also for the situation represented in the figure, as a matter of fact.

Nonetheless, in most real-world frameworks, a similar result would not be acceptable. In the case under consideration, in particular, to overcome this stumbling block, it would be sufficient to represent the empty space (internal 'hole') associated to item i as an equivalent (zero mass) additional *component* (see Fig. 2.4) and include it in the *non-intersection* conditions between items i and j.

It is understood that the additional *components* have to be neglected when not necessary, as in the case, for instance, of the items represented in Fig. 2.5. The presence of additional *components* should indeed be considered, case by case, for each pair of items, depending on the specific context under analysis. Of course, much more tricky situations could also occur, even if they are not considered here. As is easily understood, however, the approach proposed above could simply be extended, to tackle adequately the specific cases under consideration.

2.3 Additional Conditions

The MIP approach introduced in Sect. 2.1, differently from other packing methodologies (especially the ones based on sequential placement algorithms), is quite
suitable for treating a number of additional conditions. A survey of cases, quite
frequent in practice, is reported hereinafter. Before going on with this part, it is
however useful to introduce, for each pair of items i and j, the variable $\chi_{ij} \in [0, 1]$
(implicitly binary, i.e. $\chi_{ij} \in \{0, 1\}$), with the following conditions:

$$\forall i,j \in I / i < j \qquad \chi_{ij} \leq \chi_i, \tag{2.12a}$$

$$\forall i,j \in I / i < j \qquad \chi_{ij} \leq \chi_j, \tag{2.12b}$$

$$\forall i,j \in I / i < j \qquad \chi_{ij} \geq \chi_i + \chi_j - 1. \tag{2.13}$$

2.3.1 Conditions on Item Position and Orientation

Specific loading conditions, such as those concerning the pre-fixed position and
orientation of some items, are straightforward. Indeed this task, from the modelling
point of view adopted, simply consists of fixing the relative variables $o_{\beta i}$ at the desired
values and setting to one the $\vartheta_{\omega i}$ corresponding to the orientation that is targeted.

Weaker conditions on the item position can easily be introduced, if necessary, by
posing lower and upper bounds on its local reference frame origin or even on the
centre of each single *component*. Some orientations can be inhibited simply by
setting to zero the corresponding $\vartheta_{\omega i}$ this allows, for instance, to force a local
reference frame axis to be parallel to a particular direction (i.e. parallel to an axis of
the main reference frame).

Similarly, (pairwise) 'parallelism' conditions can be taken into account, by
acting properly on the relevant variables ϑ. Given, for instance, two identical
items i and j, the constraints below state that if both of them are loaded, they
must have the same orientation: $\forall \omega \in \Omega, \theta_{\omega i} \geq \theta_{\omega j} + \chi_{ij} - 1$.

2.3.2 Conditions on Pairwise Relative Distance

In some practical applications, either minimum or maximum distance conditions,
involving pairs of *components*, belonging to different items, can be posed. From a
general point of view, constraints like the following can thus be required for
components h and k of item i and j, respectively:

$$\sum_{\beta \in B} \left(w_{\beta 0 h i} - w_{\beta 0 k j} \right)^2 \geq \underline{D}^2_{hkij} \chi_{ij}, \tag{2.14}$$

$$\sum_{\beta \in B} \left(w_{\beta 0hi} - w_{\beta 0kj} \right)^2 \leq \overline{D}^2_{hkij} \chi_{ij} + \left(\sum_{\beta \in B} D^2_\beta \right) (1 - \chi_{ij}), \qquad (2.15)$$

where $\underline{D}_{hkij}$ and $\overline{D}_{hkij}$ are the given lower and upper bounds for their relative distance, respectively, whilst the other terms have already been defined. Clearly, inequality (2.14) expresses the minimum distance condition, whilst (2.15) the one on the maximum, holding if both items are loaded. Inequalities (2.14) and (2.15) become redundant, otherwise, if at least one of the items is not loaded, i.e. $\chi_{ij} = 0$.

Constraint (2.14) coincides (when $\chi_{ij} = 1$) with the classical one that guarantees the non-intersection between two spheres. This is notoriously non-convex and well known for being very difficult to deal with, as results from the specialist literature on *circle/sphere packing* (e.g. Addis et al. 2008a, b; Castillo et al. 2008; Gensane 2004, Kampas and Pintér 2005; Locatelli and Raber 2002; Stoyan and Yaskov 2008; Stoyan et al. 2003; Sutou and Dai 2002). Constraint (2.15), on the contrary, is convex and, as such, much easier to treat, even if still nonlinear (it could be observed that *tighter big-Ms* could be profitably chosen).

Remark 2.4   To show that the constraint $\sum_{\beta \in B} \left(w_{\beta 0hi} - w_{\beta 0kj} \right)^2 \leq \overline{D}^2_{hkij}$ is convex, it suffices to observe that, for any β, each quadratic form $(w_{\beta 0hi} - w_{\beta 0kj})^2$ is obviously positive semi-definite and the first member of the inequality above is thus a sum (with positive coefficients) of convex functions (e.g. Minoux and Vajda 1986).

In order to remain within an MIP framework, a piecewise linear approximation (e.g. Williams 1993) can easily be applied both to (2.14) and (2.15). To this purpose, it is useful to pose $e_{\beta hkij} = w_{\beta 0hi} - w_{\beta 0kj}$, with $e_{\beta hkij} \in [-D_\beta, D_\beta]$. Adopting an obvious simplification of the symbols, constraints (2.14) and (2.15) hence assume the form

$$\sum_{\beta \in B} e_\beta^2 \geq \underline{D}^2 \chi, \qquad (2.16)$$

$$\sum_{\beta \in B} e_\beta^2 \leq \overline{D}^2 \chi + \left(\sum_{\beta \in B} D^2_\beta \right) (1 - \chi). \qquad (2.17)$$

For the sake of simplicity, focusing just on constraint (2.16), we shall then discretize the interval $[-D_\beta, D_\beta]$ in N_S subintervals, i.e. $\left[-D_\beta, D_\beta \right] = \bigcup_{\gamma-1, \gamma \in D_s} \left[D_{\gamma-1}, D_\gamma \right]$, with $D_s = \{0, \ldots, \gamma, \ldots, N_s\}$. The terms e_β^2 can then be approximated by piecewise linear functions $([-D_\beta, D_\beta] \rightarrow [0, D_\beta^2])$ simply by posing

$$\forall \beta \in B \quad e_\beta = \sum_{\gamma \in D_s} \lambda_{\beta \gamma} D_\gamma, \qquad (2.18)$$

$$\forall \beta \in B \quad e_\beta^2 \approx \sum_{\gamma \in D_s} \lambda_{\beta \gamma} D_\gamma^2, \qquad (2.19)$$

$$\sum_{\gamma \in D_S} \lambda_{\beta\gamma} = 1, \tag{2.20}$$

where the $\lambda_{\beta\gamma}$ are non-negative variables and such that a*t most two adjacent can be non-zero (adjacency*) condition). Indeed, this restriction has to be added in order to guarantee that for each $e_\beta \in [-D_\beta, D_\beta]$, the point $\left(e_\beta, \sum_{\gamma \in D_S} \lambda_{\beta\gamma} D_\beta^2\right) \in \mathbf{R}^2$ actually lies on a segment of the corresponding piecewise linear function.

Variables subject to the *adjacency* condition stated above are said to determine a *special ordered set of variables of type 2* (SOS2, e.g. Williams 1993). It is well known that such a condition (or, more in general, *special ordered sets* of the various types) can be tackled either algorithmically (most MIP solvers have dedicated features) or by introducing additional binary variables and proper constraints in the model (an highly efficient formulation for the SOS2 case can be found in Vielma and Nemhauser 2009).

Piecewise linear approximation (and the SOS2 approach) is directly applicable to separable functions in general, but also more complex classes can be considered (e.g. Williams 1993). Here it is worth pointing out that when convex constraints (expressed by separable functions) are concerned, as, for instance, (2.17), it can easily be shown that the *adjacency* condition may be dropped, without bringing in any undesirable solution (e.g. Williams 1993). In the specific case in question, (2.18), (2.19) and (2.20) are indeed sufficient, per se, to guarantee that each point

$$\left(e_\beta, \sum_{\gamma \in \overline{D}_S} \lambda_{\beta\gamma} D_\gamma^2\right)$$ belongs to the convex domain (in $\mathbf{R}^2$) delimited by the vertices

(D_γ, D_γ^2), for all $\gamma \in D_S$. This way, the convex constraint (2.17) is always satisfied, since, for any values assumed by the variables $\lambda_{\beta\gamma}$, it is never underestimated (similar considerations hold when the minimization of a separable convex *objective* function is concerned, e.g. Williams 1993).

In addition to what is discussed above, very simple indirect conditions on the relative distance between *components* of two different items can be obtained as variations of (2.5a) and (2.5b):

$$\forall \beta \in B, \forall i,j \in I / i < j, \forall h \in C_i, \forall k \in C_j$$

$$w_{\beta 0 h i} - w_{\beta 0 k j} \geq \frac{1}{2} \sum_{\omega \in \Omega} \left(L_{\omega\beta h i}\vartheta_{\omega i} + L_{\omega\beta k j}\vartheta_{\omega j}\right) + G_{hkij} - \left(D_\beta + G_{hkij}\right)\left(1 - \sigma_{\beta hkij}^+\right),$$

$$\tag{2.21a}$$

$$\forall \beta \in B, \forall i,j \in I / i < j, \forall h \in C_i, \forall k \in C_j$$

$$w_{\beta 0 k j} - w_{\beta 0 h i} \geq \frac{1}{2} \sum_{\omega \in \Omega} \left(L_{\omega\beta h i}\vartheta_{\omega i} + L_{\omega\beta k j}\vartheta_{\omega j}\right) + G_{hkij} - \left(D_\beta + G_{hkij}\right)\left(1 - \sigma_{\beta hkij}^-\right).$$

$$\tag{2.21b}$$

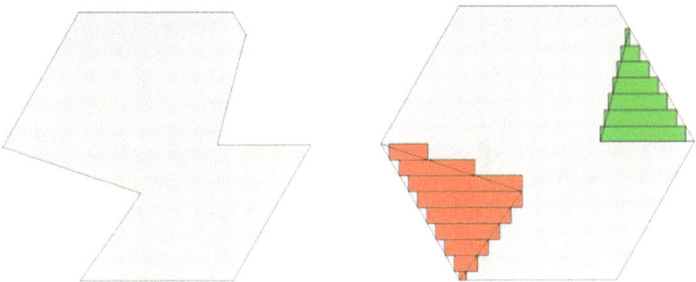

Fig. 2.6 Non-convex domain (*left*) approximated with forbidden zones (*right*, 2D representation)

Here G_{hkij} are given constants. Inequalities (2.21a) and (2.21b) thus guarantee a minimum gap between *components* h of i and k of j, respectively, along the axis corresponding to the active *non-intersection* condition. Of course, more complicated gap conditions could be stated similarly, but are not reported here.

Remark 2.5 It is, instead, worth mentioning that close-related topics come up in the context of the electronic design automation (EDA) and very large scale integration (VLSI), when dealing with the issue of minimizing the total wire length. Depending on the specific framework to look upon, the wire-length objective function terms are expressed by linear, quadratic or, more in general, nonlinear functions (e.g. Kahng and Wang 2005; Kim and Kim 2003; Kleinhans et al. 1991).

2.3.3   *Conditions on Domains*

Quite often in practice one has to take into account domains that are not convex, owing to the presence of forbidden zones, e.g. due to clearance requirements or actual 'holes'. This makes the domain non-convex, as a matter of fact.

If the forbidden zones and 'holes' are tetris-like-shaped (or properly approximated as such), they can easily be treated by the model reported in Sect. 2.1. They may indeed simply be considered as zero-mass items with given position and orientation. Analogous considerations hold also if the domain external shape is not convex, since it can easily be approximated by introducing forbidden zones, as appropriate, see Fig. 2.6.

A similar approach can be adopted, in the presence of *structural* elements that can be taken account of in terms of non-zero-mass items with fixed position and orientation: see Fig. 2.7.

Further conditions, quite useful in practice, can be introduced to deal with *separation* planes, partitioning the whole domain in sectors. They can simply be represented as 'flat' parallelepipeds: their bases are assumed to be parallel to one of the planes of the main reference frame and cover the whole domain sections they cut; their position (distance with respect to the parallel plane of the main reference frame) is allowed to vary within a given range. An alternative formulation can be applied by introducing, for each item, a set of binary variables, one per sector, that are equal to one if the item belongs to the corresponding sector and zero otherwise (Fasano 2003).

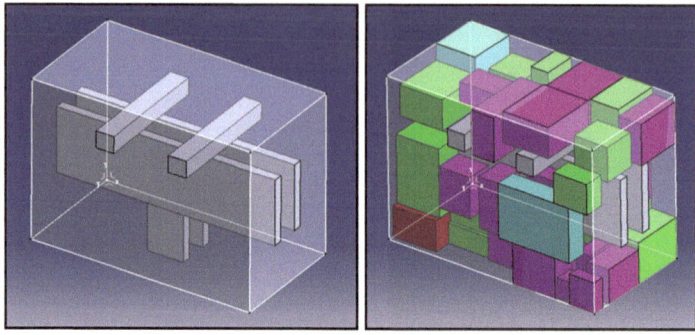

Fig. 2.7 Domain with structural elements (*left*) and items packed around (*right*)

2.3.4   Conditions on the Total Mass Loaded and Its Distribution

In several real-world applications, such as aerospace engineering and transportation systems in general, quite demanding requirements on the total mass or its distribution inside the domain have to be taken into account. Restrictions on the overall load are simply posed as follows:

$$\underline{M} \le \sum_{i \in I} M_i \chi_i \le \overline{M}, \tag{2.22}$$

where $\underline{M}$ and $\overline{M}$ are the given lower and upper mass bounds, respectively. Some insights concerning balancing conditions are provided next.

2.3.4.1   Static Balancing Restriction

The problem of loading a set of single parallelepipeds inside a convex domain, each with a given mass, so that the overall centre of mass lies within a given convex subdomain (inside the container), has been previously discussed (Fasano 2004, 2008). This restriction is denoted in the following by *static* balancing.

To generalize this issue to the case of tetris-like items, we shall firstly introduce for each one of them the terms $W^*_{\omega\beta i}$. These represent the projections, on the axes w_β, of the coordinate differences between item i centre of mass and the origin of the local reference frame, corresponding to orientations ω. With $w^*_{\beta i}$ we shall denote, for each item i, its centre of mass coordinates, with respect to the main reference frame. Conditions (2.2), (2.3) and (2.4) are then properly adapted to take into account also this point (so that, in particular, $w^*_{\beta i} = 0$ if $\chi_i = 0$). Let us indicate with D^* the convex subdomain in which the overall centre of mass must stay, in compliance with the *static* balancing restriction. Denoting with V^* the set of vertices delimiting D^* and with $V^*_{\beta\gamma}$ their coordinates, with respect to the main reference frame, the proposition stated below holds.

Proposition 2.2 *Static balancing necessary and sufficient conditions are as follows*:

$$\forall \beta \in B \quad \sum_{i \in I} M_i w^*_{\beta i} = \sum_{\gamma \in V^*} V^*_{\beta\gamma} \psi^*_\gamma, \tag{2.23}$$

$$\sum_{\gamma \in V^*} \psi^*_\gamma = m, \tag{2.24}$$

where $m = \sum_{i \in I} M_i \chi_i$ *and* $\forall \gamma \in V^* \; \psi^*_\gamma = \widetilde{\psi}^*_\gamma m$, *with* $\widetilde{\psi}^*_\gamma \geq 0$.

Proof To prove this proposition, it can simply be observed that the following conditions are necessary and sufficient for the overall centre of mass staying inside the given (sub)domain (supposing $m > 0$): $\forall \; \beta \in B \; \sum_{i \in I} \dfrac{M_i w^*_{\beta i}}{m} = \sum_{\gamma \in V^*} V^*_{\beta\gamma} \widetilde{\psi}^*_\gamma$, with $\sum_{\gamma \in V^*} \widetilde{\psi}^*_\gamma = 1$ and $\widetilde{\psi}^*_\gamma \geq 0$. Indeed they state that point $\sum_{i \in I} \dfrac{M_i w^*_{\beta i}}{m}$, i.e. the overall centre of mass, lies inside the convex (sub)domain D^* of vertices $V^*_\gamma \in V^*$. The above conditions are obviously equivalent to (2.23) and (2.24), by using $\psi^*_\gamma = m\widetilde{\psi}^*_\gamma$. (If $m = 0$, (2.23) and (2.24) are trivially satisfied, as, by (2.2), for each non-picked item i, the variables $w^*_{\beta i}$ become zero). □

Remark 2.6 It is important to point out that the correlations $\forall \; \gamma \in V^* \; \psi^*_\gamma = \widetilde{\psi}^*_\gamma m$, $\widetilde{\psi}^*_\gamma \geq 0$ can simply be substituted with $\forall \; \gamma \in V^* \; \psi^*_\gamma \geq 0$ (as the variables $\widetilde{\psi}^*_\gamma$ only play an 'ancillary' role, having no 'physical' meaning in the model). This way, conditions (2.23) and (2.24) are linear (as the nonlinear ones $\psi^*_\gamma = m\widetilde{\psi}^*_\gamma$ are omitted tout court). Moreover, the above balancing conditions are simplified when the centre of mass (sub)domain is a parallelepiped, i.e. defined, for each axis w_β, by the intervals $\left[\underline{C}^*_\beta, \overline{C}^*_\beta\right]$. In such a case, they have the simpler form $\forall \; \beta$ $\underline{C}^*_\beta m \leq \sum_{i \in I} M_i w^*_{\beta i} \leq \overline{C}^*_\beta m$. It is gathered that (2.23) and (2.24) do not take account of the mass of the container. An appropriate modification of them can easily be attained, in order to include it and its contribution to the overall centre of mass.

Remark 2.7 An item consisting of a single nonhomogeneous parallelepiped can simply be considered as composed of two elements: one parallelepiped with zero mass, geometrically identical to the item itself and its centre of mass. The composed item can hence be treated as a (degenerate) tetris-like item, whose components are the parallelepiped with zero mass and this point.

An interesting issue arises when looking upon the presence of a filling material of non-negligible density. This situation can occur, for instance, in the space engineering context, when quite dense protective foam has to fill every gap between the loaded items. To formulate the related model, it is sufficient to consider the whole domain D as if it were entirely filled with the filling material and replace all the mass associated to the volumes occupied by each item, with their actual one. Recalling the general definition of centre of mass, its coordinates w_β^* are expressed by the following equations:

$$\forall \beta \in B \quad w_\beta^* = \frac{\displaystyle\int_D u_\beta r(u) du_\beta}{m}. \tag{2.25}$$

Here m represents the total mass contained in D (referring both to the items loaded and the filling material), whilst $r(u)$ its relative density function (that in the specific case under consideration is a constant). The following equations hold:

$$\forall \beta \in B \quad mw_\beta^* = \int_{\breve{D}} u_\beta r du_\beta + \sum_{i \in I} M_i w_{\beta i}^* = \breve{M} \breve{W}_\beta^* + \sum_{i \in I} \left(M_i - \breve{M}_i \right) w_{\beta i}^*. \tag{2.26}$$

Here $\breve{D}$ is the subdomain of D corresponding to the volume not occupied by the items (and thus occupied by the filling material); $\breve{M}_i$ denotes the mass each item i would assume if its density were the same of the filling material; $\breve{M}$ and $\breve{W}^*$ indicate the total mass of the filling material and its relative centre of mass, respectively, if it filled the whole domain D (i.e. with no items inside). Thus, the following conditions extend the *static* balancing conditions (2.23) and (2.24):

$$\forall \beta \in B \quad \breve{M} \breve{W}_\beta^* + \sum_{i \in I} \left(M_i - \breve{M}_i \right) w_{\beta i}^* = \sum_{\gamma \in V^*} V_{\beta \gamma}^* \psi_\gamma^*, \tag{2.27}$$

$$\sum_{\gamma \in V^*} \psi_\gamma^* = \breve{M} - \sum_{i \in I} \left(\breve{M}_i - M_i \right) \chi_i, \tag{2.28}$$

$$\forall \gamma \in V^*, \quad \psi_\gamma^* \geq 0.$$

2.3.4.2 Dynamic Balancing Restrictions

Quite demanding requirements, involving inertia properties of the whole system, may be posed (e.g. Egeblad 2009; Limbourg et al. 2012). In space engineering, for instance, quite frequently, specific conditions on the spacecraft inertia matrix are set to address fuel consumption or attitude control concerns. Assuming the items as point masses, we shall introduce the following constraints:

$$\forall \beta, \beta' \in B / \beta < \beta'$$

$$\left| \sum_{i \in I} M_i w_{\beta i}^* w_{\beta' i}^* \right| \leq \bar{I}_{\beta \beta'}(m), \tag{2.29}$$

$$\forall \beta, \beta', \beta'' \in B/\beta < \beta', \beta, \beta' \neq \beta''$$
$$\sum_{i \in I} M_i \left(w_{\beta i}^{*2} + w_{\beta' i}^{*2} \right) \geq \underline{I}_{\beta''}(m), \tag{2.30a}$$

$$\forall \beta, \beta', \beta'' \in B/\beta < \beta', \beta, \beta' \neq \beta''$$
$$\sum_{i \in I} M_i \left(w_{\beta i}^{*2} + w_{\beta' i}^{*2} \right) \leq \bar{I}_{\beta''}(m), \tag{2.30b}$$

where $\bar{I}_{\beta\beta'}(m)$, $\underline{I}_{\beta''}(m)$ and $\bar{I}_{\beta''}(m)$ are (non-negative) functions of the total loaded mass m.

Remark 2.8 It is immediately gathered that (2.29), (2.30a) and (2.30b) are nonlinear constraints, giving rise to an MINLP model. In these constraints, the inertia characteristics of each single item have been neglected, considering, for the sake of simplicity, just simple point masses, but more precise formulations could be looked into.

It is of particular interest when $\underline{I}_1(m) \approx \bar{I}_1(m) \approx \underline{I}_2(m) \approx \bar{I}_2(m)$ (and $\bar{I}_{\beta\beta'}(m) \approx 0$). With an appropriate setting of the *static* balancing conditions, indeed, this case makes the system mass distribution assume, at a certain grade of approximation, the dynamic properties of a homogeneous cylinder.

2.3.5 Further Loading Restrictions

A significant number of further restrictions could be added, depending on the specific framework. Conditions concerning the relative position between items, such as, for instance, 'item i must stay over or under j', would also be treated easily, simply by acting properly on the relevant variables $\sigma^{+/-}$.

Much more demanding scenarios are, instead, tackled in the specialist packing literature, looking upon further additional loading conditions such as *stability*, *load bearing* and *multi-dropping* (e.g. Bischoff 2006; Bortfeldt and Gehring 2001; Christensen and Rousøe 2009; Eley 2002; Junqueira et al. 2011; Junqueira et al. 2012; Lai et al. 1998; Morabito and Arenales 1994; Moura and Oliveira 2005; Pisinger 2002; Ratcliff and Bischoff 1998; Silva et al. 2003).

These can be defined as follows (Junqueira et al. 2013): 'Cargo *stability* refers to the support of the bottom faces of boxes, in the case of vertical *stability* (i.e., the boxes must have their bottom faces supported by other box top faces or the container floor), and the support of the lateral faces of boxes, in the case of horizontal *stability*. *Load bearing* strength refers to the maximum number of boxes that can be stacked one above each other, or more generally, to the maximum pressure that can be applied over the top face of a box. *Multi-dropping* refers to cases where boxes that are delivered to the same customer (destination) must be placed close to each other inside the container and the loading pattern must take into account the delivery route of the vehicle and the sequence in which the boxes are unloaded'.

Dealing with such additional conditions, an interesting approach has been suggested (Junqueira et al. 2013). It develops a 'grid-based position paradigm' (i.e. *space-indexed* formulation, e.g. Allen et al. 2012), as opposed to the 'position free' one, i.e. the MIP model of Sect. 2.1 (or equivalent versions), that instead allows for continuous item positions.

Extensions of the single parallelepiped MIP model (Sect. 2.1), aimed at tackling a number of additional conditions, are taken into account by Pesch (working paper). Those, for instance, denoted by *layer* constraints consider the presence of incompatibility relations of the type: item i cannot be positioned 'on the top' of item j. As pointed out in this work, they can be represented by a directed graph G, where $arc(i, j) \in G$ if and only if item i cannot be placed on item j. Moreover, if an item is not placed on the floor, it has to be supported by at least another item.

Whilst such additional conditions can be expressed by an MINLP formulation (Pesch, working paper), an alternative MIP model is briefly discussed here, also referring, for the sake of simplicity, to the case of single parallelepipeds.

To this purpose, we shall introduce, firstly, the binary variables $\underline{\chi}_i$ and $\widehat{\chi}_i$, with the meaning:

$\underline{\chi}_i = 1$ if item i lies on the container basis; $\underline{\chi}_i = 0$ either if it is not loaded or it is supported by (at least) another item.

$\widehat{\chi}_i = 1$ if item i is supported by (at least) another item; $\widehat{\chi}_i = 0$ either if it is not loaded or it lies on the container basis.

These have the task of controlling the status of possible contact, for each item i, with respect to the lower basis of the container (floor). They are linked, in a mutually exclusive mode, to the corresponding variables χ_i as follows:

$$\forall i \in I \quad \underline{\chi}_i + \widehat{\chi}_i = \chi_i,$$

This way, if item i is picked, then one and only one of the two related statuses is admissible. Thus, assuming that the axis w_3 of the main reference frame is the 'vertical' one, the conditions below hold:

$$\forall i \in I \quad w_{3i} \geq \frac{1}{2} \sum_{\alpha \in A} L_{\alpha i} \delta_{\alpha 3i} - \frac{1}{2} L_{3i} \left(1 - \underline{\chi}_i \right),$$

$$\forall i \in I \quad w_{3i} \leq \frac{1}{2} \sum_{\alpha \in A} L_{\alpha i} \delta_{\alpha 3i} + (D_3 - L_{1i}) \left(1 - \underline{\chi}_i \right).$$

These constraints force item i to lie on the container base, when the corresponding condition is active, i.e. $\underline{\chi}_i = 1$, and become redundant otherwise, i.e. when $\underline{\chi}_i = 0$.

In addition to this, the case corresponding to $\widehat{\chi}_i = 1$, i.e. when item i is picked but it is supported by (at least) another item, has to be properly examined. Before going ahead with this point, however, we shall introduce the binary variables $\widehat{\sigma}^+_{3ij}$ and $\widehat{\sigma}^-_{3ij}$, with the meaning:

$\widehat{\sigma}^+_{3ij} = 1$ if item i is placed on the top of j and zero otherwise.

$\widehat{\sigma}^-_{3ij} = 1$ if item j is placed on the top of i and zero otherwise.

The inequalities below are thus added to (2.10a) and (2.10b), respectively:

$$\forall i, j \in I / i < j \quad w_{3i} - w_{3j} \geq \frac{1}{2} \sum_{\alpha \in A} \left(L_{\alpha i} \delta_{\alpha 3i} + L_{\alpha j} \delta_{\alpha 3j} \right) - \left(1 - \widehat{\sigma}^+_{3ij} \right) D_3,$$

$$\forall i, j \in I / i < j \quad w_{3i} - w_{3j} \leq \frac{1}{2} \sum_{\alpha \in A} \left(L_{\alpha i} \delta_{\alpha 3i} + L_{\alpha j} \delta_{\alpha 3j} \right) + \left(1 - \widehat{\sigma}^+_{3ij} \right) D_3,$$

$$\forall i, j \in I / i < j \quad w_{3j} - w_{3i} \geq \frac{1}{2} \sum_{\alpha \in A} \left(L_{\alpha i} \delta_{\alpha 3i} + L_{\alpha j} \delta_{\alpha 3j} \right) - \left(1 - \widehat{\sigma}^-_{3ij} \right) D_3,$$

$$\forall i, j \in I / i < j \quad w_{3j} - w_{3i} \leq \frac{1}{2} \sum_{\alpha \in A} \left(L_{\alpha i} \delta_{\alpha 3i} + L_{\alpha j} \delta_{\alpha 3j} \right) + \left(1 - \widehat{\sigma}^-_{3ij} \right) D_3.$$

They imply, as is immediately seen, that $w_{3i} = w_{3j} + \frac{1}{2} \sum_{\alpha \in A} \left(L_{\alpha i} \delta_{\alpha 3i} + L_{\alpha j} \delta_{\alpha 3j} \right)$ (i.e. i lies on the top of j), when $\widehat{\sigma}^+_{3ij} = 1$, and $w_{3j} = w_{3i} + \frac{1}{2} \sum_{\alpha \in A} \left(L_{\alpha i} \delta_{\alpha 3i} + L_{\alpha j} \delta_{\alpha 3j} \right)$ (i.e. j lies on the top of i), when $\widehat{\sigma}^-_{3ij} = 1$. The first pair of inequalities becomes redundant otherwise when $\widehat{\sigma}^+_{3ij} = 0$ and, similarly for the second one, when $\widehat{\sigma}^-_{3ij} = 0$. Inequalities (2.11) are then extended as follows:

$$\forall i, j \in I / i < j \quad \sum_{\beta \in B} \left(\sigma^+_{\beta ij} + \sigma^-_{\beta ij} \right) + \widehat{\sigma}^+_{3ij} + \widehat{\sigma}^-_{3ij} \geq \chi_i + \chi_j - 1,$$

where the terms $\widehat{\sigma}^+_{3ij}$ are set, a priori, to zero if $\text{arc}(i, j) \in G$ and, analogously, for $\widehat{\sigma}^-_{3ij}$, if $\text{arc}(j, i) \in G$. The conditions stated below imply that if $\widehat{\chi}_i = 1$, item i has to be positioned on the top of (at least) another one:

$$\forall i \in I \quad \widehat{\chi}_i \leq \sum_{\substack{j \in I / \\ i < j}} \widehat{\sigma}^+_{3ij} + \sum_{\substack{i' \in I / \\ i' < i}} \widehat{\sigma}^-_{3i'i}.$$

Further conditions can be taken into account to include the *stability* requirements (e.g. Pesch, working paper). This way, it is guaranteed that when item i is on the top of item j, the projection of its centre of mass, on the 'horizontal' plane, i.e. (O, w_1, w_2), lies inside the rectangle determined by the projection of item j. The following conditions (or more refined ones, with *tighter big-Ms*) can thus be added:

$$\beta \in \{1,2\}, \forall i,j \in I/i < j \quad w_{\beta i} \leq w_{\beta j} + \frac{1}{2}\sum_{\alpha \in A} L_{\alpha j}\delta_{\alpha \beta j} + \left(1 - \widehat{\sigma}^{+}_{3ij}\right)D_{\beta},$$

$$\beta \in \{1,2\}, \forall i,j \in I/i < j \quad w_{\beta i} \geq w_{\beta j} - \frac{1}{2}\sum_{\alpha \in A} L_{\alpha j}\delta_{\alpha \beta j} - \left(1 - \widehat{\sigma}^{+}_{3ij}\right)D_{\beta}.$$

The *layer* and *stability* additional conditions can efficiently be tackled algorithmically, by means of a dedicated heuristic (Pesch, working paper). The (non-trivial) computational aspects, related to the extended MIP model discussed above (or possible alternative formulations), could, nevertheless, represent an interesting line for further in-depth investigation.

Chapter 3
Model Reformulations and Tightening

The general MIP model, discussed in Chap. 2, is reconsidered hereinafter, investigating some possible reformulations, from different points of view (Sect. 3.1). The objective of enucleating implicit implications and introducing *valid* inequalities, to *tighten* the model, is examined next (Sect. 3.2).

3.1 Alternative Models

The issue discussed in this section focuses mainly on the case occurring when the packing problem is expressed in terms of *feasibility*, i.e. when all the given items have to be placed and no *objective* function is stated a priori. This situation can arise, for instance, when the items are the elements of a device and, as such, they all have to be installed inside an appropriate container, as essential parts of the same equipment. The thus defined *feasibility* subproblem is also of interest, as it represents one of the basic concepts of the heuristic procedures put forward in Chap. 4. As far as this specific subproblem is concerned, since no *objective* function is specified a priori, an arbitrary one can be introduced, in order to simplify the task of finding an *integer-feasible* solution.

The general model of Sect. 2.1 (including the additional conditions of Sect. 2.3) is reconsidered hereinafter in terms of *feasibility*, providing three different reformulations of it (Sects. 3.1.1, 3.1.2 and 3.1.4). In all of them, it is understood that either all the given items can be loaded or the instance to solve is infeasible. In each of these reformulations, an ad hoc *objective* function is defined, with the scope of minimizing (even if indirectly) the overall overlap of items. In the first (Sect. 3.1.1) and second (Sect. 3.1.2, except the variation outlined at the end), no sooner does the solver obtain the first *integer-feasible* solution than the optimization is stopped (even if just a suboptimal solution of the ad hoc *objective* function has been found). In all reformulations, both the *orthogonality* and *domain* conditions are maintained, as defined in Sect. 2.1. (i.e. consisting of constraints (2.1), (2.2), (2.3) and (2.4)). The second reformulation (Sect. 3.1.2) is subject to

G. Fasano, *Solving Non-standard Packing Problems by Global Optimization and Heuristics*, SpringerBriefs in Optimization, DOI 10.1007/978-3-319-05005-8_3, © Giorgio Fasano 2014

straightforward variations. One in particular (Sect. 3.1.3) is an actual alternative to the general MIP model, no longer restricted to the *feasibility* subproblem. It could also be utilized (at least partially) in the heuristics of Chap. 4. This aspect would definitely represent an interesting objective for future research.

3.1.1   General MIP Model First Linear Reformulation

The rationale of the general MIP model reformulation presented hereinafter stresses the introduction of an ad hoc *objective* function. This aims at reducing the solution search region, as much as possible, in order to obtain any *integer-feasible* solution.

The approach adopted draws on the work achieved by Suhl (1984), dealing with (large-scale) *fixed-charge* models. Suhl's work provides an efficient preprocessing technique aimed at reducing the *big-M* terms, associated to the *fixed-charge* constraints, i.e. at 'minimizing' (a priori) the related region, in the LP *relaxation*.

As far as the model reformulation in question is concerned, an approach, intended to 'minimize' the search region R_S, relative to the *non-intersection* (*big-M*) constraints (2.5a) and (2.5b), is investigated, to tackle efficiently the relative *feasibility* subproblem. These constraints are then reformulated in an *LP-relaxed* form and an ad hoc *objective* function, substituting (2.7), is introduced. The reformulated model is described as follows.

All variables χ are set to one, as all the given items must be inside the domain and the *non-intersection* constraints (2.5a) and (2.5b) are rewritten as

$$\forall \beta \in B, \forall i,j \in I / i < j, \forall h \in C_i, \forall k \in C_j$$

$$w_{\beta 0hi} - w_{\beta 0kj} \geq \frac{1}{2} \sum_{\omega \in \Omega} \left(L_{\omega \beta hi} \vartheta_{\omega i} + L_{\omega \beta kj} \vartheta_{\omega j} \right) + d^+_{\beta hkij} - D_\beta, \tag{3.1a}$$

$$\forall \beta \in B, \forall i,j \in I / i < j, \forall h \in C_i, \forall k \in C_j$$

$$w_{\beta 0kj} - w_{\beta 0hi} \geq \frac{1}{2} \sum_{\omega \in \Omega} \left(L_{\omega \beta hi} \vartheta_{\omega i} + L_{\omega \beta kj} \vartheta_{\omega j} \right) + d^-_{\beta hkij} - D_\beta. \tag{3.1b}$$

Constraints (2.6) are substituted with the following:

$$\forall \beta \in B, \forall i,j \in I / i < j, \forall h \in C_i, \forall k \in C_j \quad d^+_{\beta hkij} \geq \sigma^+_{\beta hkij} D_\beta, \tag{3.2a}$$

$$\forall \beta \in B, \forall i,j \in I / i < j, \forall h \in C_i, \forall k \in C_j \quad d^-_{\beta hkij} \geq \sigma^-_{\beta hkij} D_\beta, \tag{3.2b}$$

$$\forall i,j \in I / i < j, \forall h \in C_i, \forall k \in C_j$$

$$\sum_{\beta \in B} \left(\sigma^+_{\beta hkij} + \sigma^-_{\beta hkij} \right) = 1, \tag{3.3}$$

where $d^+_{\beta hkij}, d^-_{\beta hkij} \in [0, D_\beta]$.

The adopted ad hoc *objective* function is

$$\max \sum_{\substack{\beta \in B, \\ i,j \in I/i < j, \\ h \in C_i, k \in C_j}} \frac{d^+_{\beta hkij} + d^-_{\beta hkij}}{D_\beta}. \tag{3.4}$$

Any optimal solution of the reformulated model identifies a minimal subset of the *feasibility* region, relative to the general MIP model (Sect. 2.1).

Proposition 3.1 *For any given set of items, the feasibility regions, associated to the general MIP model and its first linear reformulation respectively (neglecting the subspace associated to the variables d^+ and d^-), are coincident.*

Proof Dealing with the feasibility subproblem, all variables χ are set to one. Constraints (2.1), (2.2), (2.3) and (2.4) are obviously coincident in both models, and it is thus sufficient to demonstrate that constraints (2.5a), (2.5b) and (2.6) of the general MIP model are equivalent to constraints (3.1a), (3.1b), (3.2a), (3.2b) and (3.3) of the reformulated one. It is immediately seen that given that all variables χ are set to one, constraints (2.6) can be substituted with (3.3). To show that constraints (2.5a) and (2.5b) are equivalent to (3.1a), (3.1b), (3.2a) and (3.2b), we shall distinguish the cases where the variables σ are zero from those where they are equal to one.

Consider, for instance, $\sigma^+_{\beta hkij} = 0$. This implies that constraints (2.5a) become

$$w_{\beta 0hi} - w_{\beta 0kj} \geq \frac{1}{2} \sum_{\omega \in \Omega} \left(L_{\omega\beta hi}\vartheta_{\omega i} + L_{\omega\beta kj}\vartheta_{\omega j} \right) - D_\beta.$$

These are equivalent to constraints (3.1a), with $d^+_{\beta hkij} = 0$, in compliance with constraints (3.2a). Considering, instead, $\sigma^+_{\beta hkij} = 1$, this implies that constraints

(2.5a) become $w_{\beta 0hi} - w_{\beta 0kj} \geq \frac{1}{2} \sum_{\omega \in \Omega} \left(L_{\omega\beta hi}\vartheta_{\omega i} + L_{\omega\beta kj}\vartheta_{\omega j} \right)$.

These are equivalent to constraints (3.1a), with $d^+_{\beta hkij} = D_\beta$, in compliance with constraints (3.2a). As the same reasoning can be carried out, taking into account the cases relative to the variables $\sigma^-_{\beta hkij}$, the two models are equivalent. □

Remark 3.1 To better understand the meaning of the general MIP model first linear reformulation, we shall make some intuitive considerations. Let us define, for each β, for every pair of components h and k of item i and j, respectively, the squared subspace $S_\beta = [0, D_\beta] \times [0, D_\beta] \subset R^2$, associated to variables $d^-_{\beta hkij}$ and $d^+_{\beta hkij}$. The bound $d^+_{\beta hkij} + d^-_{\beta hkij} \leq 2D_\beta - \sum_{\omega \in \Omega} \left(L_{\omega\beta hi}\vartheta_{\omega i} + L_{\omega\beta kj}\vartheta_{\omega j} \right)$ is implicitly determined by inequalities (3.1a) and (3.1b). The objective function induces the solution projection on S_β to stay along the straight line $d^+_{\beta hkij} + d^-_{\beta hkij} = 2D_\beta - \sum_{\omega \in \Omega} \left(L_{\omega\beta hi}\vartheta_{\omega i} + L_{\omega\beta kj}\vartheta_{\omega j} \right)$.

If $\quad D_\beta - \sum_{\omega \in \Omega} \left(L_{\omega\beta hi}\vartheta_{\omega i} + L_{\omega\beta kj}\vartheta_{\omega j} \right) \geq 0, \quad$ this $\quad$ intersects $\quad S_\beta \quad$ in $\quad$ the $\quad$ points

$$\left(D_\beta, D_\beta - \sum_{\omega \in \Omega}\left(L_{\omega\beta hi}\vartheta_{\omega i} + L_{\omega\beta kj}\vartheta_{\omega j}\right)\right) \text{ and } \left(D_\beta - \sum_{\omega \in \Omega}\left(L_{\omega\beta hi}\vartheta_{\omega i} + L_{\omega\beta kj}\vartheta_{\omega j}\right), D_\beta\right),$$

respectively, determining an internal segment. In this occurrence, if the linear solver (utilized by the MIP optimizer) looks for vertex solutions (as in the case of a simplex-based one), the above extreme points are more likely to be selected than the ones internal to the segment (although this expectation is not based on rigorous reasoning). One has to bear in mind, moreover, that either $d^+_{\beta hkij} = D_\beta$ or $d^-_{\beta hkij} = D_\beta$ (for any β) guarantees that no intersection occurs between the two corresponding items.

As a partially alternative version of this model reformulation, the constraints $\forall \beta \in B, \forall i, j \in I/i < j, \forall h \in C_i, \forall k \in C_j, d^+_{\beta hkij} + d^-_{\beta hkij} \leq D_\beta$ could also be added to *tighten* the *feasibility* region (creating in the subspace S_β the two extreme points $(D_\beta, 0)$ and $(0, D_\beta)$, without excluding any solution. These inequalities are obviously *tighter* than the bounds $d^+_{\beta hkij} + d^-_{\beta hkij} \leq 2D_\beta - \sum_{\omega \in \Omega}\left(L_{\omega\beta hi}\vartheta_{\omega i} + L_{\omega\beta kj}\vartheta_{\omega j}\right)$,

when $D_\beta - \sum_{\omega \in \Omega}\left(L_{\omega\beta hi}\vartheta_{\omega i} + L_{\omega\beta kj}\vartheta_{\omega j}\right) \geq 0$. The conditions $d^-_{\beta hkij}, d^+_{\beta hkij} \in [0, D_\beta]$, moreover, if explicitly introduced in the model, can be of computational advantage, when the linear solver adopted treats the variable bounds independently (as in the case of *simplex-based* ones).

3.1.2 General MIP Model Second Linear Reformulation

To discuss this alternative model, we shall consider, for each item *component*, the set of all concentric parallelepipeds containing it. The reformulation examined hereinafter is also based on an ad hoc *objective* function. It is aimed at finding, for each *component*, the enclosing parallelepiped (included in D) of maximum volume that does not intersect any other enclosing parallelepipeds, associated to *components* of different items.

To this purpose, the *non-intersection* conditions of Sect. 2.1 are properly changed. Whilst (2.6) is kept, inequalities (2.5a) and (2.5b) are substituted with the constraints below. For each *component* h of i, the non-negative variables $l_{\beta hi}$ are introduced, assuming that all variables χ are set to one:

$$\forall \beta \in B, \forall i, j \in I/i < j, \forall h \in C_i, \forall k \in C_j$$

$$w_{\beta 0hi} - w_{\beta 0kj} \geq \frac{1}{2}\left(l_{\beta hi} + l_{\beta kj}\right) - D_\beta\left(1 - \sigma^+_{\beta hkij}\right), \tag{3.5a}$$

$$\forall \beta \in B, \forall i, j \in I/i < j, \forall h \in C_i, \forall k \in C_j$$

$$w_{\beta 0kj} - w_{\beta 0hi} \geq \frac{1}{2}\left(l_{\beta hi} + l_{\beta kj}\right) - D_\beta\left(1 - \sigma^-_{\beta hkij}\right), \tag{3.5b}$$

$$\forall \omega \in \Omega, \forall \beta \in B, \forall i \in I, \forall h \in C_i$$

$$l_{\beta hi} \geq L_{\omega \beta hi} \vartheta_{\omega i}. \tag{3.6}$$

The following (*surrogate*) *objective* function is defined:

$$\max \sum_{\substack{\beta \in B, \\ i \in I, h \in C_i}} l_{\beta hi}. \tag{3.7}$$

For each *component h* of each item *i*, the terms $l_{\beta hi}$ represent (for the orientation ω assumed by *i*) the projections, on the axes w_β, of an enclosing parallelepiped, containing *component h* and centred with it. Inequalities (3.5a), (3.5b) and (3.6) (together with (2.6)) guarantee that the enclosing parallelepipeds, belonging to different items, do not intersect.

Remark 3.2 Rigorously speaking, as the objective function (3.7) refers to the total sum of the component sides, it should be considered as a surrogate expression of $\max \sum_{i \in I, h \in C_i} \prod_{\beta \in B} l_{\beta hi}.$

As previously mentioned, possible variations of the approach discussed above could be considered. One is obtained simply by inverting inequalities (3.6) as follows and keeping all remaining constraints, as well as the *objective* function, unaltered:

$$\forall \omega \in \Omega, \forall \beta \in B, \forall i \in I, \forall h \in C_i$$

$$l_{\beta hi} \leq L_{\omega \beta hi} \vartheta_{\omega i}. \tag{3.8}$$

In this case, an *integer-optimal* solution (and not just any *integer-feasible* one) has necessarily to be found, in order to guarantee that no intersections occur among the given items. It should be noticed that, at each step, the optimization process is induced to minimize the overall overlap, without assigning items a volume that exceeds their own. Moreover, since, in this case, the value of the global optimal solution is known a priori, it can be advantageously utilized as *cutoff* parameter (to get rid of suboptimal solutions).

3.1.3   A Non-restrictive Reformulation of the General MIP Model

A possible reformulation of the general MIP model, without renouncing its original objective of maximizing either the overall loaded volume or mass, is also quite straightforward. The problem is no longer expressed in terms of *feasibility* (i.e. without the possibility of rejecting items), so that all variables χ are set free again, as in Sect. 2.1.

As a first step, inequalities (3.6) are transformed into the equations:

$$\forall \omega \in \Omega, \forall \beta \in B, \forall i \in I, \forall h \in C_i$$

$$l_{\beta h i} = L_{\omega \beta h i} \vartheta_{\omega i}. \tag{3.9}$$

In order to define the new *objective* function (substituting (2.7)), the terms K_{hi} are introduced (with obvious meaning) for each *component* h of each item i, where $\forall i \in I \sum_{h \in C_i} K_{hi} = K_i$, cf. (2.7). The dimensions of *component* h of i are indicated with $L_{\alpha h i}$, $\alpha \in \{1, 2, 3\} = A$, assuming, from now on, that $L_{1hi} \leq L_{2hi} \leq L_{3hi}$. The new *objective* function is then expressed by the following:

$$\max \sum_{\substack{\beta \in B, \\ i \in I, h \in C_i}} \frac{K_{hi}}{\sum_{\alpha \in A} L_{\alpha h i}} l_{\beta h i}. \tag{3.10}$$

It is easily seen that the two *objective* functions (2.7) and (3.10) are equivalent for any *integer-feasible* solution (by (3.9)). The expression (3.10), differently from (2.7), provides the significant computational advantage of minimizing the item overall overlap at each step of the optimization process. Just to summarize the reformulation in question, we could point out that it consists of constraints (2.1), (2.2) (*orthogonality*), (2.3), (2.4) (*domain*), (2.6), (3.5a), (3.5b) and (3.9) (*non-intersection*), in addition to *objective* function (3.10). It is also understood that in all the relevant expressions above, the variables $l_{\beta h i}$ could be eliminated. They may, indeed, be substituted by their corresponding terms, on the basis of (3.9) (that could also be eliminated).

3.1.4 General MIP Model Nonlinear Reformulation

The general packing problem presented in Sect. 2.1 is notoriously subject to nonlinear (MINLP) formulations (e.g. Birgin and Lobato 2010; Birgin et al. 2006; Cassioli and Locatelli 2011). We shall introduce, hereinafter, a nonlinear reformulation of the general MIP model *non-intersection* constraints, assuming, as previously, that all variables χ are set to one (as the *feasibility* subproblem is in question). It is straightforward to prove that the nonlinear constraints below are equivalent to (2.5a), (2.5b) and (2.6):

$\forall \beta \in B, \forall i,j \in I / i < j, \forall h \in C_i, \forall k \in C_j$

$$\left(w_{\beta 0 h i} - w_{\beta 0 k j}\right)^2 - \left[\frac{1}{2} \sum_{\omega \in \Omega} \left(L_{\omega \beta h i} \vartheta_{\omega i} + L_{\omega \beta k j} \vartheta_{\omega j}\right)\right]^2 = s_{\beta h k i j} - r_{\beta h k i j}, \tag{3.11}$$

$\forall \beta \in B, \forall i,j \in I / i < j, \forall h \in C_i, \forall k \in C_j$

$$\prod_{\beta \in B} r_{\beta h k i j} = 0, \tag{3.12}$$

where $s_{\beta h k i j} \in [0, D_\beta^2]$ and $r_{\beta h k i j} \in [0, D_\beta^2]$ (actually, smaller upper bounds could be chosen for both sets of variables).

Indeed, for each pair of *components* h and k, of items i and j, respectively, equations (3.12) guarantee that for at least one β, the corresponding term $r_{\beta h k i j}$ is zero, and equations (3.11) that the *non-intersection* conditions hold for such a β, i.e. $\left|w_{\beta 0 h i} - w_{\beta 0 k j}\right| \geq \frac{1}{2} \sum_{\omega \in \Omega} \left(L_{\omega \beta h i} \vartheta_{\omega i} + L_{\omega \beta k j} \vartheta_{\omega j}\right)$. More precisely, constraints (2.5a) and (2.5b) correspond to equations (3.11), whilst equations (2.6) correspond to (3.12).

As the *non-intersection* constraints (3.11) and (3.12) are most likely hard to tackle, they are therefore considered in terms of *fixed penalization* in the ad hoc *objective* function we are going to introduce. All remaining linear (MIP), constraints are kept as such. A formulation aimed at satisfying as much *non-intersection* conditions as possible is the following:

$$\min \left\{ \sum_{\substack{\beta \in B, \\ i,j \in I / i < j, \\ h \in C_i, k \in C_j}} \left\{ \left(w_{\beta 0 h i} - w_{\beta 0 k j}\right)^2 - \left[\frac{1}{2} \sum_{\omega \in \Omega} \left(L_{\omega \beta h i} \vartheta_{\omega i} + L_{\omega \beta k j} \vartheta_{\omega j}\right)\right]^2 - s_{\beta h k i j} + r_{\beta h k i j} \right\}^2 \right.$$

$$\left. + K_P \sum_{\substack{i,j \in I / i < j, \\ h \in C_i, k \in C_j}} \prod_{\beta \in B} r_{\beta h k i j} \right\}$$

$$\tag{3.13}$$

where K_P is a positive coefficient (that represents an appropriate 'weight' associated to the product terms).

It is immediately seen that the *objective* function (3.13) is non-negative. A zero-global-optimal solution exists if and only if the constraints ((2.1), (2.2), (2.3), (2.4), (2.5a), (2.5b) and (2.6) of the general MIP model of Sect. 2.1 (with all variables χ set to one) delimit a feasible region. This *objective* function thus 'minimizes' the intersection between items. Its global optima, moreover, guarantee an ultimate (non-approximate) solution to the *feasibility* subproblem under discussion. It could be observed that for each set of variables ϑ, (3.13) is a polynomial function (providing, as such, potential algorithmic advantages; on global polynomial optimization, see, for instance, De Loera et al. 2012; Hanzon and Jibetean 2003; Schweighofer 2006).

Alternative *fixed penalization* can be considered (e.g. Cassioli and Locatelli 2011). We shall introduce here one *objective function* with *fixed penalization* correlated to the *non-intersection* constraints only:

$$\min\left\{ \sum_{\substack{\beta \in B, \\ i,j \in I/i < j, \\ h \in C_i, k \in C_j}} \max\left\{ -\left(w_{\beta 0 h i} - w_{\beta 0 k j}\right)^2 + \left[\frac{1}{2}\sum_{\omega \in \Omega}\left(L_{\omega\beta h i}\vartheta_{\omega i} + L_{\omega\beta k j}\vartheta_{\omega j}\right)\right]^2, 0 \right\} \right.$$

$$\left. + K_P \sum_{\substack{i,j \in I/i < j, \\ h \in C_i, k \in C_j,}} \prod_{\beta \in B} r_{\beta h k i j} \right\}$$

$$(3.14)$$

As the previous one, this *objective* function is also non-negative and each zero-global-optimum corresponds to a solution of the *feasibility* problem.

Remark 3.3 Both the MINLP formulations discussed above contain only linear (MIP) constraints. This aspect could be advantageous, when the MINLP solvers utilized treat the model linear sub-structure independently (e.g. The MathWorks 2012). It is moreover worth noticing that all functions involved in both MINLP formulations are Lipschitz-continuous and, consequently, Lipschitzian solvers can be profitably adopted (e.g. Pintér 1997, 2009). Indeed, all constraints are of the MIP type and (3.13) is smooth. As far as (3.14) is concerned, it is sufficient to observe

that the terms $\max\left\{ -\left(w_{\beta 0 h i} - w_{\beta 0 k j}\right)^2 + \left[\frac{1}{2}\sum_{\omega \in \Omega}\left(L_{\omega\beta h i}\vartheta_{\omega i} + L_{\omega\beta k j}\vartheta_{\omega j}\right)\right]^2, 0 \right\}$ keep

their Lipshitz-continuous characteristic (e.g. Pintér 1996).

3.2 Implications and Valid Inequalities

As is well known, in the MIP context, remarkable research effort has been devoted to looking into general approaches to *tighten* the model. This means to make its *linear relaxation* an as precise as possible approximation of the *convex hull* relative to the *mixed-integer* solutions (e.g. Andersen et al. 2005; Ceria et al. 1998; De Farias et al. 1998; Jünger et al. 2009; Marchand et al 1999; Nemhauser and Wolsey 1990; Van Roy and Wolsey 1987; Weismantel 1996; Wolsey 1989).

Polyhedral analysis (e.g. Atamtürk 2005; Constantino 1998; Dash et al. 2010; Hamacher et al. 2004; Padberg 1995; Pochet and Wolsey 1994; Yaman 2009) is adopted to this purpose, in order to find *valid* inequalities (e.g. Aardal et al. 1995; Cornuéjols 2008; Padberg et al. 1985; Wolsey 1990, 2003). These are aimed at *tightening* the MIP model under consideration. The introduction of such auxiliary conditions is particularly suitable when a *branch-and-cut* approach (e.g. Andreello et al. 2007; Balas et al. 1996; Cordier et al. 2001; Padberg 2001; Padberg and Rinaldi 1991) is followed.

Differently from more traditional MIP algorithms, such as *branch-and-bound* (where all model constraints have to be set a priori) with a *branch-and-cut* process, the *valid* inequalities are activated just when needed and dropped when not required.

With reference to the general MIP model (Sect. 2.1), for items consisting of single parallelepipeds to load into a parallelepiped (see Sect. 2.1, special case), some *valid* inequalities, holding under specific assumptions, have been put forward by Padberg (1999). This has been done to tackle the problem by means of a dedicated *branch-and-cut* approach. Some quite simple conditions, not restricted to the case of single parallelepipeds, are considered hereinafter (limited subsets of them can be advantageously taken into account also when a *branch-and-bound* approach is adopted). A first group of inequalities is hence introduced:

$$\forall i,j \in I/i < j, \forall h \in C_i, \forall k \in C_j$$

$$\sum_{\beta \in B} \left(\sigma^+_{\beta hkij} + \sigma^-_{\beta hkij} \right) \leq \chi_i, \tag{3.15a}$$

$$\forall i,j \in I/i < j, \forall h \in C_i, \forall k \in C_j$$

$$\sum_{\beta \in B} \left(\sigma^+_{\beta hkij} + \sigma^-_{\beta hkij} \right) \leq \chi_j. \tag{3.15b}$$

These, together with (2.6), for each pair of *components* h and k of items i and j, respectively, imply that one, and only one, of the relative variables $\sigma^+_{\beta hkij}$ and $\sigma^-_{\beta hkij}$ has to be equal to one if both items are loaded; all of them are equal to zero otherwise. It is immediate to notice that in the general MIP model of Sect. 2.1, in case both items are picked, more than one of the variables $\sigma^+_{\beta hkij}$ and $\sigma^-_{\beta hkij}$ could be non-zero. The above extended version is hence *tighter* than the previous, without any loss of generality, as no *integer-feasible* solutions are excluded.

Some straightforward examples of necessary conditions, concerning pairs of items, in particular situations, can be considered. Firstly, let us consider the very simple case when item i and j cannot be aligned with respect to the axis w_β (because they would exceed the dimension D_β, for all possible orientations of both). In such an occurrence, the conditions below can be explicitly posed:

$$\forall h \in C_i, \forall k \in C_j \quad \sigma^+_{\beta hkij} = \sigma^-_{\beta hkij} = 0.$$

In addition to these, a set of more complicated implications, correlating alignment and orientation, could be introduced. An example, dealing with the special case of Sect. 2.1, relative to single parallelepipeds, is reported here.[1] Considering items i and j, if $L_{1i} + L_{2j} > D_\beta$, *they cannot be aligned along the axis* w_β, *with either* L_{2j} *or* L_{3j} *parallel to it.* And analogously, this holds if $L_{1j} + L_{2i} > D_\beta$. The following inequalities can hence be set:

$$\forall \beta, \forall i, j \in I / i < j, L_{1i} + L_{2j} > D_\beta \quad \delta_{2\beta j} + \delta_{3\beta j} \le 1 - \sigma^+_{\beta ij} - \sigma^-_{\beta ij},$$

$$\forall \beta, \forall i, j \in I / i < j, L_{1j} + L_{2i} > D_\beta \quad \delta_{2\beta i} + \delta_{3\beta i} \le 1 - \sigma^+_{\beta ij} - \sigma^-_{\beta ij}.$$

These conditions can easily be extended when tetris-like items are involved, i.e. when the general MIP model of Sect. 2.1 is considered. This gives rise to inequalities of the type $\displaystyle\sum_{\omega \in \Omega'_{\beta hkij}} \vartheta_{\omega j} \le 1 - \sigma^+_{\beta hkij} - \sigma^-_{\beta hkij}$, where $\Omega'_{\beta hkij}$ $(i < j)$ is the set of orientations (of j), incompatible with the alignment conditions of the components h (of i) and k (of j). Similar expressions hold for i, with $\Omega'_{\beta hkji}$ $(i < j)$.

Straightforward *transitivity* conditions (e.g. Padberg 1999; Fasano 2008) can, moreover, be looked upon, when triplets of single parallelepipeds are taken into account. They can easily be extended when actual tetris-like items are involved. Focusing on the triplet of *components* h, h', h'' of items i, i', i'', respectively, *if, along the axis* w_β, h *precedes* h' *and* h' *precedes* h'', *then* h *precedes* h'', *along the same axis.* This implication is expressed by

$$\forall \beta \in B, \forall i, i', i'' \in I / i < i' < i'', \forall h \in C_i, \forall h' \in C_{i'}, \forall h'' \in C_{i''}$$

$$\sigma^-_{\beta h h'' i i''} \ge \sigma^-_{\beta h h' i i'} + \sigma^-_{\beta h' h'' i' i''} - 1.$$

Still referring to the same triplet of *components*, the further implication holds: *if* $L_{1hi} + L_{1h'i'} + L_{1h''i''} > D_\beta$, *then the whole triplet cannot be aligned along the axis* w_β. This is expressed by the following constraints:

$$\forall \beta \in B, \forall i, i', i'' \in I / i < i' < i'', \forall h \in C_i, \forall h' \in C_{i'}, \forall h'' \in C_{i''} / L_{1hi} + L_{1h'i'} + L_{1h''i''} > D_\beta$$

$$\sigma^+_{\beta h h' i i'} + \sigma^-_{\beta h h' i i'} + \sigma^+_{\beta h' h'' i' i''} + \sigma^-_{\beta h' h'' i' i''} + \sigma^+_{\beta h h'' i i''} + \sigma^-_{\beta h h'' i i''} \le 2.$$

The proof is straightforward. It is sufficient to notice that being the hypothesis stated, at the most, two *components* may be aligned along the axis w_β and that for each pair of *components*, either the corresponding variable σ^+_β or σ^-_β must be zero.

---

[1] Note These conditions have been introduced by S. Gliozzi, senior managing consultant at IBM GBS Advanced Analytics and Optimization.

As a further observation, note that the implications correlating alignment and orientation, as presented in this section, would be susceptible to extensions involving chains of more than three *components*. Their introduction could provide practical advantages in the perspective of a dedicated *branch-and-cut* approach.

Remark 3.4 When the layer constraints reported in Sect. 2.3.5 are introduced in the model, inequalities (3.15a) and (3.15b) can properly be extended. Moreover, the necessary conditions $\forall i \in I \ w_{3i} \geq \min_{i' \neq i} \{L_{1i'}\} \widehat{\chi}_i$ can explicitly be added, following the perspective presented in this section.

Chapter 4
Heuristic Approaches for Solving the Tetris-like Item Problem in Practice

As easily gathered, the general MIP model, conceived to sort out the *tetris*-like item packing problem (Sect. 2.1), is usually very hard to solve. In this chapter the relevant intrinsic difficulties are examined first (Sect. 4.1). A heuristic philosophy is then emphasized to tackle efficiently the problem, even if just nonproven optimal solutions can, in general, be obtained.

The basic concept of *abstract configuration* is introduced (Sect. 4.2). Chapter 3 reformulations, devised to solve the *feasibility* subproblem, are exploited to look into an initial approximate solution (Sect. 4.3.1). Two alternative heuristic procedures, thought up to improve it recursively, until a satisfactory result is reached, are discussed next (Sects. 4.3.2 and 4.3.3). The possibility of interacting with the solution process is also outlined (Sect. 4.3.4).

4.1 Intrinsic Difficulties

Broad classes of packing problems are notoriously well known for being *NP-hard* (as regards this classification and, more in general, the theoretical aspects related to *computational complexity* and *approximability*, see, for instance, Ausiello et al. 2003, Chlebík and Chlebíková 2006, Goldreich 2008).

The general MIP model reported in Chap. 2 is, per se, extremely hard to solve in practice, when real-world instances have to be dealt with. This holds, even if only single parallelepipeds are involved and no additional conditions are set (see Sect. 2.1, special case). In this circumstance, insights on its complexity can be provided by looking upon the model overall structure (N indicates here the total number of single parallelepipeds):

$O(3N(N-1))$ binary variables σ
$O(9N)$ binary variables δ
$O(N)$ binary variables χ
$O(6N)$ *orthogonality* constraints

G. Fasano, *Solving Non-standard Packing Problems by Global Optimization and Heuristics*, SpringerBriefs in Optimization, DOI 10.1007/978-3-319-05005-8_4, © Giorgio Fasano 2014

$O(3N)$ *domain* constraints

$$O\left(7\frac{N(N-1)}{2}\right) \; \textit{non-intersection constraints (of which } O(3N(N-1))\textit{big-Ms)}$$

The model scale increases quite dramatically when N actual tetris-like items are involved. In order to discuss this case, we shall indicate, for each item i, with $|C_i|$ the cardinality of the set C_i relative to its *components*. The total number of pairs of *components*, belonging to different items, is denoted by N_C and computed as follows:

$$N_C = \left(\frac{\sum_{i \in I} |C_i|}{2}\right) - \sum_{i \in I}\left(\frac{|C_i|}{2}\right), \tag{4.1}$$

where the symbol $\begin{pmatrix} N_1 \\ N_2 \end{pmatrix}$ represents, in general, the N_2-combinations of a set containing N_1-elements. The model overall structure is hence represented by the following:

$O(24N)$ binary variables θ
$O(N)$ binary variables χ
$O(6N_C)$ binary variables σ
$O(7N_C)$ *non-intersection* constraints (of which $O(6N_C)$ *big-Ms*).

Even at a first glance, the intricacy of both scenarios mentioned above is not only related to the binary variables. Indeed, it primarily depends on the presence of the *big-M* constraints, related to the *non-intersection* conditions. The occurrence of the implicit *transitivity* conditions (see Sect. 3.2), moreover, provides a significant insight on the hidden model difficulties. The following proposition shows how their total number can be computed.

Proposition 4.1 *Given N tetris-like items, the total number of transitivity conditions is* $18 \sum\limits_{\substack{i,i',i'' \in I/ \\ i < i' < i''}} |C_i| \cdot |C_{i'}| \cdot |C_{i''}|$.

Proof To prove this proposition we can firstly concentrate on the particular case of Sect. 3.2, where the sequence h precedes h' and h' precedes h'' ($h \prec h' \prec h''$) along the (general) axis w_β is considered. The conditions $\forall \beta \in B$, $\forall i, i', i'' \in I/i < i' < i''$, $\forall h \in C_i$, $\forall h' \in C_{i'}$, $\forall h'' \in C_{i''} \sigma^-_{\beta h h'' i i''} \geq \sigma^-_{\beta h h' i i'} + \sigma^-_{\beta h' h'' i' i''} - 1$ are therefore recalled.

Selecting (for the time being) the specific axis w_β, for a given pair of components h and h' of i and i', respectively, there are $|C_{i''}|$ such conditions; for a given component h of i, they are $|C_{i'}| \cdot |C_{i''}|$, so that their total number is $|C_i| \cdot |C_{i'}| \cdot |C_{i''}|$. This corresponds to the order relation $(h \in C_i) \prec (h' \in C_{i'}) \prec (h'' \in C_{i''})$. There are $3!$ permutations of such a kind, for which the above transitivity conditions have to be properly rearranged (utilizing appropriately the variables σ^- and σ^+). This must hold for any triplet i, i' and i'', such that $i < i' < i''$, and for each axis w_β. $\square$

Remark 4.1 When all the tetris-like items involved have the same number ($|C_i|$, $\forall\ i\ \in\ I$) of components, the total number of transitivity conditions is $18|C_i|^3\binom{N}{3}$.

   To consider a quantitative example, an instance involving 50 items, of five *components* each, the order of magnitude relevant to all the *non-intersection* (*big-M*) constraints is $6 \times 3 \times 10^4$; that of the *transitivity* conditions is 4×10^7. Section 4.2 investigates the sets of variables σ^+ and σ^- (at most one for each pair of components belonging to different items) that, if fixed to one, are compliant with the *transitivity* implications. These variables σ are called *transitivity compatible*.

4.2   Abstract Configurations

The *abstract configuration* concept is introduced by the following definitions.

Definition 4.1   Constraints of the types

$$w_{\beta0hi} - w_{\beta0kj} \geq \frac{1}{2}\sum_{\omega \in \Omega}\left(L_{\omega\beta hi}\vartheta_{\omega i} + L_{\omega\beta kj}\vartheta_{\omega j}\right),$$

$$w_{\beta0kj} - w_{\beta0hi} \geq \frac{1}{2}\sum_{\omega \in \Omega}\left(L_{\omega\beta hi}\vartheta_{\omega i} + L_{\omega\beta kj}\vartheta_{\omega j}\right),$$

corresponding to either $\sigma^+_{\beta hkij} = 1$ or $\sigma^-_{\beta hkij} = 1$ in (2.5a) and (2.5b), respectively, are called relative position constraint*s*.

Definition 4.2   Given *a* set of N items and all the N_C pairs of components, belonging to different items, an abstract configuration consists of N_C relative position constraint*s* (one and only one for each pair) compatible in any unbounded domain.

Remark 4.2 From Definition 4.2, it follows immediately that, for a given set of N items, an abstract configuration corresponds to a set of N_C transitivity-compatible σ variables fixed to one.

   To interpret the concept of *abstract configuration*, it is opportune to restrict the discussion to the simpler case of single-*component* items (see Sect. 2.1, special case). Theoretically, in such a case, all the *abstract configurations*, associated to a given set of N items, could be directly derived by adopting a three-dimensional generalization of the *square grid graph* (e.g. Weisstein 2012). This is the graph whose vertices correspond to the points of $N^3 \subset R^3$ (referred to an orthogonal reference frame), with integer coordinates that are in the range $0, \ldots, N - 1$ and with any two vertices connected by an edge, whenever they are at unit distance; see Fig. 4.1.

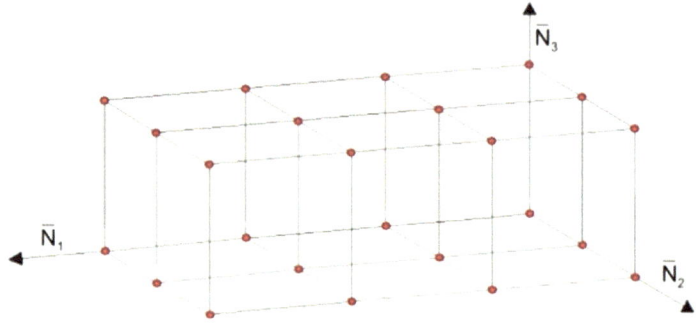

Fig. 4.1 Three-dimensional grid

Since the *relative positions* are to be considered in an unbounded domain, from this 'topological' perspective, the actual dimensions of the items can be totally neglected. Items can, thus, simply be considered as geometrical points and their relative placement represented by N nodes (intended as points of N^3) of the aforementioned grid. A set of corresponding *relative position* constraints can hence be selected to generate an *abstract configuration*. If, for instance, item i is associated to node $(0,0,0)$ and item j is associated to node $(N-1, N-1, N-1)$, their coordinate relative distance on the grid, with respect to each axis, is $(N-1)$ units and i precedes j in all directions. This, in an unbounded domain, is compliant, with the following *relative position* constraints:

$$w_{1j} - w_{1i} \geq \frac{1}{2} \sum_{\alpha \in A} \left(L_{\alpha i} \delta_{\alpha 1 i} + L_{\alpha j} \delta_{\alpha 1 j} \right),$$

$$w_{2j} - w_{2i} \geq \frac{1}{2} \sum_{\alpha \in A} \left(L_{\alpha i} \delta_{\alpha 2 i} + L_{\alpha j} \delta_{\alpha 2 j} \right),$$

$$w_{3j} - w_{3i} \geq \frac{1}{2} \sum_{\alpha \in A} \left(L_{\alpha i} \delta_{\alpha 3 i} + L_{\alpha j} \delta_{\alpha 3 j} \right).$$

Any of the above *relative position* constraints can thus be selected, as the one corresponding to the pair of items (i,j), to create an *abstract configuration* (relative to the given set of N items). Similar considerations hold for all the possible $\binom{N^3}{N}$ associations of N items to the N^3 grid nodes. It is worth noticing that, in any unbounded domain, each given *abstract configuration* yields an infinity of packing scenarios, obtainable by simple roto-translations of the items. This suggests that even when an *abstract configuration* is forced, if compliant with the given domain, the items still have a certain freedom of movement.

As it is immediately understood, even for relatively small-scale instances, the number of all possible associations of N items to N^3 grid nodes is immense.

Of course, it is higher than that of the *abstract configurations* which can actually be generated, as the following example shows. Let us consider the set of three items i, i' and i'', together with the two following distinct associations A_1 and A_2:

$$A_1) \quad i \to (0,0,0), \quad i' \to (2,0,0), \quad i'' \to (1,2,0);$$
$$A_2) \quad i \to (0,1,0), \quad i' \to (2,1,0), \quad i'' \to (1,2,0).$$

These are compliant with the same *relative position* constraints and, thus, give rise to the same corresponding *abstract configurations*.

Considering also that some associations 'dominate' others, in the sense that their corresponding sets of *abstract configurations* strictly contain those of the 'dominated' ones, several further redundancies are expected. It is gathered that quite a lot of duplications could be eliminated. The number of grid nodes could be reduced, for instance, taking into account the actual size of the domain D. Indeed (see Fig. 4.1), its maximum value, on each axis w_β, respectively, is given by

$$\forall \beta \in B \quad \overline{N}_\beta = \max_{\substack{N' \leq N \\ \sum_{i=1}^{N'} L_{1i} \leq D_\beta}} \left\{ N' \right\}.$$

It is, nevertheless, obvious that any exhaustive approach, based on the generation of all possible *abstract configurations,* would be impracticable, in real-world cases. It is, moreover, quite intuitive that a generalized use of the three-dimensional grid, to cope also with the case of tetris-like items (and not only of single parallelepipeds), would be quite tricky indeed.

The basic idea of the heuristic approaches put forward in Sect. 4.3 addresses instead the exploitation of a number of 'good' *abstract configurations*. The following discussion focuses therefore on the capability of extracting, from approximate solutions (with possible item overlap), *abstract configurations* that, at least partially, consider the actual characteristics of the problem (e.g. conditions on items, domain and balancing). The thus obtained *abstract configurations* are then exploited to give rise to *integer-feasible* (even if, generally, suboptimal) solutions, compliant with all the given conditions (including, most certainly, *orthogonality*, *domain* and *non-intersection*).

Given any approximate solution, the objective of the *abstract configuration* generation consists of assigning one and only one *relative position* constraint to each pair of *components*, belonging to different items (in the following, this is understood, when referring to any pair of *components*).

The (arbitrary) rules listed below represent very simple selection criteria the *abstract configuration* generation could be based on; see Fig. 4.2:

- If the *components* do not intersect and the *non-intersection* conditions hold only with respect to one axis, the corresponding *relative position* constraint is selected.

Fig. 4.2 *Abstract configuration* derived from an approximate solution (two-dimensional representation of single-component items)

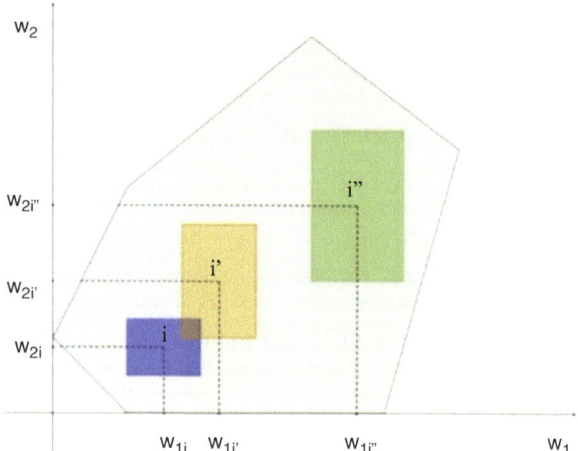

- If the *components* do not intersect and the *non-intersection* conditions hold for more than one axis, the *relative position* corresponding to the maximum distance between the projections of the respective coordinate centres is selected.
- If the *components* intersect, the *relative position* corresponding to the maximum distance between the projections of the respective coordinate centres is selected.

Figure 4.2 provides a (two-dimensional) representation of the above selection rules. Just single-*component* items, denoted as i, i' and i'', respectively, are considered, to make the example clearer (the generalization to actual tetris-like items is straightforward). As easily seen in the figure, they are subject to the following conditions:

- Items i and i' projections overlap on both the axes w_1 and w_2, with $w_{2i'} - w_{2i} > w_{1i'} - w_{1i}$.
- Items i and i'' projections neither overlap on the axis w_1 nor on w_2, with $w_{1i''} - w_{1i} > w_{2i''} - w_{2i}$.
- Items i' and i'' projections do not overlap on the axis w_1 only, with $w_{1i''} > w_{1i'}$.

Based on the rules listed above, the following *abstract configuration* (relative to the single-*component* items i, i' and i'') is extracted:

$$w_{2i'} - w_{2i} \geq \frac{1}{2} \sum_{\alpha \in A} \left(L_{\alpha i} \delta_{\alpha 2i} + L_{\alpha j} \delta_{\alpha 2i'} \right),$$

$$w_{1i''} - w_{1i} \geq \frac{1}{2} \sum_{\alpha \in A} \left(L_{\alpha i} \delta_{\alpha 1i} + L_{\alpha j} \delta_{\alpha 1i''} \right),$$

$$w_{1i''} - w_{1i'} \geq \frac{1}{2} \sum_{\alpha \in A} \left(L_{\alpha i} \delta_{\alpha 1i} + L_{\alpha j} \delta_{\alpha 1i''} \right).$$

Remark 4.3   As the above selection criteria are not based on rigorous reasoning, it is evident that a number of different (and supposedly more sophisticated) empirical rules could be explored. In case of intersection, the third rule proposed above could be, for instance, substituted with the one that selects the relative position corresponding to the projection where the minimum overlap occurs.

4.3   Solution Search

As already pointed out, the generation of *abstract configurations*, on the basis of approximate solutions, represents a fundamental concept of the overall philosophy adopted here. The first step of both the heuristic processes that are to be addressed in this section focuses, therefore, on the creation of a high-quality starting approximate solution. It is trusted, indeed, that the 'closer' to an actual solution it is, the 'better' the generated *abstract configuration* results. Consequently, less computational effort is needed to obtain a satisfactory ultimate result. This step is denoted as *initialization*.

The heuristic processes outlined in Sects. 4.3.2 and 4.3.3, respectively, provide, as a matter of fact, two alternative search strategies to work out (at least at a suboptimal level) the general MIP model (with possible additional conditions). As an overall rule of thumb, the first is more oriented to solving quite tricky instances but with a relatively limited total number (<100) of item *components* involved. The second is mainly proposed for larger ones. A joint use of the two (also in a parallelized mode) could be subject to further investigation.

4.3.1   Initialization

The reformulated models of Sect. 3.1, aimed at solving the *feasibility* subproblem, can properly be adapted to serve the *initialization* purpose. Partial *LP relaxations*, either of the first (Sect. 3.1.1) or the second (Sect. 3.1.2) linear reformulations (including all the given additional conditions or a subset of them), are hence utilized. They have the scope of finding a first approximate solution, to be refined (if opportune) by the nonlinear reformulation (Sect. 3.1.4).

In all such approximate solutions, the overlap of items is allowed, but it is minimized, in the sense specified, for each reformulated version, in Sect. 3.1. On the contrary, both the *orthogonality* and *domain* constraints are imposed. Of course, the more the additional conditions are included (even if in approximate versions), the more the obtained starting solution is a realistic representation of the actual problem to solve.

Hereinafter, we shall reexamine, in this perspective, the three reformulations, one at a time. In all these cases, we shall still consider the problem in terms of *feasibility*, as it will be tentatively assumed, a priori, that all the given items can be loaded.

This way, all variables χ are set to one and the constraints involving them readjusted properly. In each reformulation, as stated above, constraints (2.1), (2.2), (2.3) and (2.4) are maintained. In the following discussion, for the sake of simplicity, the additional conditions are neglected, as it is understood that they can be properly added each time, accordingly to the specific problem to deal with.

4.3.1.1   Use of the First Linear Reformulation

A partial *LP relaxation* of the first linear reformulation is carried out by eliminating inequalities (3.2a), (3.2b) and (3.3), whilst keeping (3.1a), (3.1b) and the *objective* function (3.4) unaltered.

Remark 4.4 It should be observed that, even when no additional conditions are present, the (partially) *LP*-relax*ed* model is still of an MIP type. This is due to the presence of the orientation variables. When the number of components involved is very high, the orientation of some (or all) items could be pre-fixed (on the basis of some empirical criterion). Analogously, it could be done for the biggest side.

4.3.1.2   Use of the Second Linear Reformulation

The second linear reformulation can be adopted, as an alternative to the first. The *non-intersection* constraints (3.5a) and (3.5b) are kept the same, whilst, as partial *LP relaxation*, inequalities (3.6) are dropped. The *objective* function (3.7) is maintained as is. This way, whilst the model is significantly simplified, it is no longer guaranteed that each component is actually enclosed by the associated parallelepiped (whose volume is maximized, in the sense specified in Remark 3.2).

If, moreover, all *components*, once reduced to the corresponding cubes of sides L_{1hi}, can actually be loaded into D, the following bounds are advantageously imposed. They are aimed at mitigating the mutual competition among items, i.e. avoiding that excessive volume is attributed to some *components* to the detriment of others:

$$\forall \beta \in B, \forall i \in I, \forall h \in C_i \quad L_{1hi} \leq l_{\beta hi} \leq L_{3hi}. \tag{4.2}$$

Otherwise, if not all the cubes of sides L_{1hi} can be loaded, the lower bounds appearing in (4.2) may be properly reduced, for instance, by subsequent attempts. Alternatively, they could be rewritten as $L_{1hi} - r_{\beta hi} \leq l_{\beta hi}$, where the terms $r_{\beta hi}$ are non-negative variables. In such a case, the objective function (3.7) can be substituted with $\max \sum_{\substack{\beta \in B, \\ i \in I, h \in C_i}} \left(l_{\beta hi} - K_R r_{\beta hi} \right)$, where K_R represents an appropriate (positive) coefficient.

It should, moreover, be noticed that the approximate solution obtained by the second linear reformulation, when the (lower) bounds (4.2) are introduced, provides a set of *transitivity-compatible* variables σ (relative to all the items involved). Since they already represent an *abstract configuration*, no further generation of it is needed.

If the model variation outlined at the end of Sect. 3.1.2 (substituting (3.6) with (3.8)) is to be applied, a possible 'relaxation' can simply be obtained by renouncing the global optimal solution (cf. Sect. 4.3.1.3) and thus admitting possible intersections among items (quite satisfactory suboptimal solutions are expected to be found with a moderate computational effort).

4.3.1.3   Use of the Nonlinear Reformulation

As already mentioned, this model is aimed at improving the approximate solutions obtained either by the first or second reformulations. The MIP solution obtained by either the first or second linear reformulation is then inherited to initialize its related Mixed Integer Non-Linear Programming (MINLP) search process.

In this case, the 'relaxation' just concerns the solution quality, in the sense that no global optimal solution is requested. Looking for local optimal solutions only, indeed, means trying to minimize the item overall overlap, but, obviously, it is not guaranteed that the *non-intersection* constraints are satisfied.

Remark 4.5 A possible alternative to the above nonlinear reformulation consists of substituting, in the first linear one, the objective function (3.4), with the following:

$$\max \sum_{\substack{\beta \in B, \\ i,j \in I/i < j, \\ h \in C_i, k \in C_j}} \frac{\left(d^+_{\beta hkij} - d^-_{\beta hkij}\right)^2}{D^2_\beta}.$$

This way, the item overall intersection is no longer minimized with direct 'competition' between each pair of variables $d^+_{\beta hkij}$ and $d^-_{\beta hkij}$, relative to the same reference frame axis w_β. Nonetheless, it should be noticed that, with this formulation, even the attainment of a global optimum is not up to guaranteeing that no intersection occurs.

Remark 4.6 It should be noticed that the total superposition of the centres of two components can occur both with the first linear reformulation and the second one, when the (lower) bounds (4.2) are neglected. In this case, the approximate solution can properly be perturbed, in order to get rid of this ambiguity (with the rules presented, for instance, in Sect. 4.2, to generate the abstract configuration, indeed, in case of such superposition, no relative position could be selected). This aspect is, however, not considered here, for the sake of simplicity.

Remark 4.7 It is gathered that, as a first (and quite daring) attempt, the abstract configuration derived from the initialization step could be directly forced into the general MIP model, requesting that all items are loaded. To do this, it would be sufficient to substitute all the non-intersection constraints (2.5a), (2.5b) and (2.6) with the available relative position ones. It is obvious, however, that an integer-feasible solution could hardly be found this way, even for very simple instances.

Remark 4.8 As some items may have prefixed orientation, this aspect could be taken into account also during the initialization step (within the limits implied by its characteristic of admitting overlap). When utilizing the first linear reformulation or the nonlinear one, the imposition of the given pre-orientations is straightforward. The situation is more complicated when the second linear reformulation is adopted. In such a case, the related model can be properly modified.

Let us suppose, for the sake of simplicity, that all items are pre-oriented. Inequalities (3.5a) and (3.5b) can be modified as follows:

$$\forall \beta \in B, \forall i, j \in I / i < j, \forall h \in C_i, \forall k \in C_j$$

$$w_{\beta 0 h i} - w_{\beta 0 k j} \geq \frac{1}{2}\left(L'_{\beta h i} - l'_{\beta h i} + L'_{\beta k j} - l'_{\beta k j}\right) - D_\beta\left(1 - \sigma^+_{\beta h k i j}\right),$$

$$\forall \beta \in B, \forall i, j \in I / i < j, \forall h \in C_i, \forall k \in C_j$$

$$w_{\beta 0 k j} - w_{\beta 0 h i} \geq \frac{1}{2}\left(L'_{\beta h i} - l'_{\beta h i} + L'_{\beta k j} - l'_{\beta k j}\right) - D_\beta\left(1 - \sigma^-_{\beta h k i j}\right).$$

Here, for each *component* h of item i, the terms $L'_{\beta h i}$ represent its pre-oriented sides, and $l'_{\beta h i}$ are non-negative variables. The *objective* function $\min \sum\limits_{\substack{\beta \in B, \\ i \in I, h \in C_i}} l'_{\beta h i}$ can then substitute (3.7) and bounds (4.2) have to be properly rewritten.

Remark 4.9 When the number of items/components involved is large, the initialization step, independently from the formulations adopted, may require quite a significant computational effort. To cope with this practical difficulty, the items involved in the original instance can be partitioned in subsets and added incrementally. At each step, the abstract configuration corresponding to the items already loaded is imposed (taking appropriate precautions to prevent possible infeasibilities). Once a new subset of items has been added, the corresponding solution gives rise to an upgraded abstract configuration, relevant to all the so-far-loaded items.

Notice that a similar recursive approach could advantageously be adopted by utilizing the non-restrictive reformulation of Sect. 3.1.3. Whenever it stops yielding improvements, either the first or the second linear reformulations can take its place in the process, in order to eventually include all the items of the original instance. It is understood that the use of the non-restrictive reformulation could be quite advantageous, since, for each (suboptimal) solution found, no overlap can occur.

4.3.2 Heuristic Process Based on Suggested Abstract Configurations

The heuristic process outlined here stresses the idea of inheriting the *abstract configuration* derived from the *initialization* step (Sects. 4.2 and 4.3.1). It is exploited (by subsequent adaptations) to obtain a satisfactory solution to the general MIP model (inclusive of the additional conditions if present, see Sect. 2.3). The *abstract configuration* and its modifications, achieved during the whole procedure, are 'suggested' recursively, adding or subtracting, time after time, items from the given set. This is carried out by means of a dichotomous approach.

In order to 'suggest' any *abstract configuration*, the general MIP model *objective* function (2.7) has to be properly modified. For this purpose, let us firstly recall that, as pointed out in Remark 4.2, any *abstract configuration* can be represented by a set of (*transitivity-compatible*) variables σ^+ and σ^-, fixed to one. This way, on the basis of the *abstract configuration* that is to be 'suggested', the following coefficients $\hat{\sigma}^+_{\beta hkij}$ and $\hat{\sigma}^-_{\beta hkij} \in \{0, 1\}$ are introduced:

$$\forall \beta \in B, \forall i,j \in I/i < j, \forall h \in C_i, \forall k \in C_j$$

$\hat{\sigma}^+_{\beta hkij} = 1$ if the *relative position* constraint $w_{\beta 0hi} - w_{\beta 0kj} \geq \frac{1}{2} \sum_{\omega \in \Omega} \left(L_{\omega \beta hi} \ \vartheta_{\omega i} \ + L_{\omega \beta kj} \vartheta_{\omega j} \right)$ belongs to the *abstract configuration*; $\hat{\sigma}^+_{\beta hkij} = 0$ otherwise.

Analogous constraints can be stated for $\hat{\sigma}^-_{\beta hkij}$. The *objective* function below substitutes then (2.7):

$$\max_{\substack{\beta \in B, \\ i,j \in I/i < j, \\ h \in C_i, k \in C_j}} \sum \left(\hat{\sigma}^+_{\beta hkij} \sigma^+_{\beta hkij} + \hat{\sigma}^-_{\beta hkij} \sigma^-_{\beta hkij} \right). \tag{4.3}$$

As is easily realized, this new optimization criterion has the effect of inducing each variable σ, corresponding to a non-zero $\hat{\sigma}$, to attain the value of one, in compliance with the 'suggested' *abstract configuration*. This has the expected effect of obtaining an *integer-feasible* solution, when existing, by a dramatically reduced computational effort.

In order to perform the heuristic process under consideration here, a first attempt is made by imposing that all the given items must be loaded. The *abstract configuration* derived from the *initialization* step is thus adopted. If any *integer-feasible* solution to the general MIP model, modified with the ad hoc *objective* function (4.3), is obtained, the given problem is solved. In the opposite case, the following dichotomous procedure is started.

An initial subset of all the given items is taken, by imposing (still tentatively) that all of them must be loaded. If any *integer-feasible* solution is obtained, a new *abstract configuration* (relative to the whole set of items) is derived, simply by carrying out a new *initialization* step: for all the items already loaded, their *relative positions* are fixed (when the nonlinear reformulation is used, the model includes them, as explicit constraints). A new *abstract configuration* is then generated.

In the opposite case, i.e. when the above subset of items does not allow for any *integer-feasible* solution, some of them are rejected, until a successful result is obtained. The heuristic process is then executed by activating, recursively, either the forward or the backward steps just described. Since the process tries and modifies, time after time, the 'suggested' *abstract configurations* as little as possible, the whole effect is that of performing an overall *depth-first* strategy.

Remark 4.10 Different versions of the heuristic approach described in this section could of course be considered. When 'suggesting' the current abstract configuration, for instance, some of the corresponding relative position constraints could be tentatively forced.

4.3.3   Heuristic Process Based on Imposed Abstract Configurations

A heuristic procedure, based on the imposition of *abstract configurations* , has been previously introduced (Fasano 2008) to tackle instances involving only single parallelepipeds to load into a convex domain. The adopted approach has been refined and extended to the case of actual tetris-like items, as outlined hereinafter.

This process, summed up by Fig. 4.3, is aimed at generating and imposing a sequence of 'good' *abstract configurations* and at solving correspondingly, step by step, a reduced MIP model, until a final satisfactory solution is attained.

The reduced model (inclusive of the additional conditions, when present; see Sect. 2.3) is derived, each time, from the general one (see Sect. 2.1) by eliminating all the redundant *non-intersection* constraints and variables σ, not contemplated by the *abstract configuration* imposed.

The *packing* module (see Fig. 4.3) is assigned the task of solving, time after time, the reduced models. Items are rejected, if necessary, to make the current *abstract configuration* (referred to the whole set of the given items) compatible with the given domain D (and the additional conditions, if any). The *item-exchange* and *hole-filling* modules (see Fig. 4.3) are employed, during the whole process, to provide new approximate solutions and supposedly improved *abstract configurations*. The *abstract configurations* are generated (on the basis of what is discussed in Sect. 4.2) from the approximate solutions obtained either by the *initialization* (Sect. 4.3.1) or the *hole-filling* steps. When derived from this one, the *relative positions* of the items previously loaded by the *packing* module are contemplated. The *item-exchange* step directly provides a new *abstract configuration*, so that no

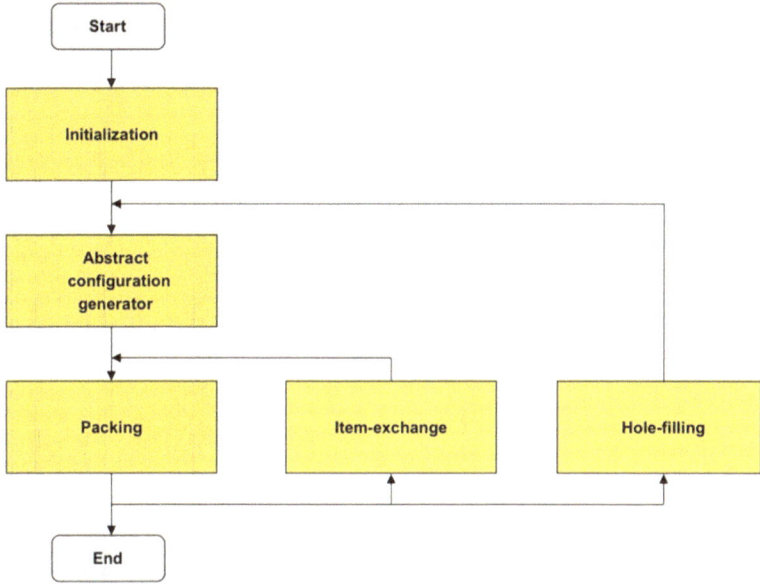

Fig. 4.3 Heuristic process based on the *abstract configuration* imposition

proper generation activity has to be performed. The *packing* module is discussed hereinafter, together with those dedicated to the *item-exchange* and *hole-filling* steps, neglecting several details, unnecessary for an overall comprehension. The general methodology is further outlined, suggesting a possible alternative approach.

4.3.3.1   Packing

The task of this module is that of obtaining an *integer-feasible* solution to the general MIP model (with possible additional conditions, see Sects. 2.1 and 2.3) by imposing an *abstract configuration*. The *non-intersection* constraints (2.5a) and (2.5b), corresponding to the *relative positions* of the *abstract configuration* imposed, are kept unaltered, with their associated variables σ. All the remaining ones are instead eliminated. Inequalities (2.6) are reduced, for all the variables σ involved, to the following:

$$\forall i, j \in I/i < j, \forall h \in C_i, \forall k \in C_j \quad \sigma_{\beta h k i j}^{+/-} \geq \chi_i + \chi_j - 1. \tag{4.4}$$

All other constraints and variables of the general MIP model (including possible additional conditions) are kept and the *objective* function (2.7) still has the purpose of maximizing either the overall loaded volume or mass.

As an interesting alternative, the non-restrictive reformulation (Sect. 3.1.3), with all the relevant constraints, inclusive of possible additional conditions, and the *objective* function (3.10) could be used. The imposition of the *abstract configurations* would be carried out as explained above, involving the appropriate constraints.

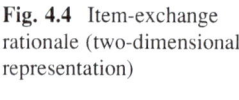

Fig. 4.4 Item-exchange rationale (two-dimensional representation)

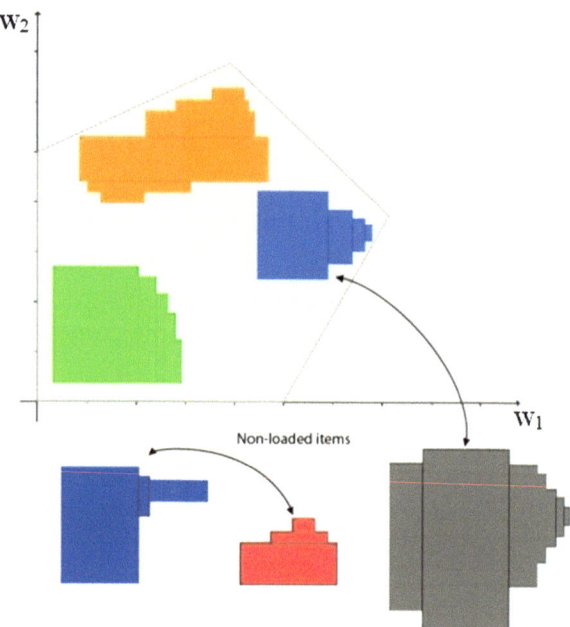

Remark 4.11 It can immediately be observed that because of constraints (4.4), the integrality condition on the variables σ may be dropped, so that they can simply be considered as continuous in the interval [0,1]. Furthermore, for each pair of components, one and only one of the non-intersection constraints (2.5a) and (2.5b) is still present, leading to a dramatic reduction of the original model dimension.

4.3.3.2   Item Exchange

This module is aimed at perturbing and tentatively improving the *abstract config-uration* (referred to the whole set of items), corresponding to the current *packing* solution. An index permutation p among the set of items (and consequently in all the *relative position* constraints) is thus executed: $\forall\ i \in I,\ i \rightarrow p(i)$. In such a way, the overall effect consists of exchanging some of their relative positions, within the given *abstract configuration*, providing a new one.

In order to carry out (at least supposedly) promising exchanges, the following heuristic rationale, open to different possible versions, is proposed. It is sketched in Fig. 4.4. Picked items are exchanged with bigger non-picked items (or with items with bigger mass, if this corresponds to the optimization criterion chosen). Non-loaded items can also be exchanged. The way the above exchanges are implemented determines the specific strategy followed.

Remark 4.12 The item-exchange module performs actions likely to be advanta-geous in terms of loaded volume (or mass), just by performing permutations. It does not take into account (directly) any constraints of the general MIP model that are, instead, contemplated by the packing module and (partially) by the hole-filling one.

Remark 4.13 Depending on the strategy adopted, this module, even by exchanging a limited number of items, can accomplish either a 'weak' or a 'strong' perturbation of the current abstract configuration. When a 'weak' perturbation strategy is executed, the exchanged items are not too different (in terms of volume and/or mass) from each other. They are, on the contrary, quite different, when a 'strong' perturbation strategy is chosen. When a 'weak' one is followed, the new abstract configuration remains 'close' to the previous, and the same is expected for the corresponding solution. This way, the general MIP model constraints (and the additional ones, when present) are therefore indirectly considered, through the 'neighbourhood' with the previous solutions .

Remark 4.14 If not all the exchanges carried out, in the new abstract configuration, between selected and nonselected items are feasible, the packing module is forced to actuate some rejections. It is, however, possible to avoid this inconvenience. Indeed, let us consider, for instance, the potential exchange of the picked item i with the non-picked one i'. It would be sufficient to duplicate, in the current packing MIP model, for item i', all relative position constraints corresponding to i and pose the further condition: $\chi_i + \chi_{i'} = 1$ (updating, subsequently, the abstract configuration, on the basis of the obtained solution). This way the relevant exchange would not be imposed (preventing the possible consequence of rejection of both items).

4.3.3.3 Hole Filling

Also this module is aimed at perturbing the *packing* module (current) solution. Empty spaces are exploited by tentatively adding items extracted from the set $\hat{I}_E$ of the currently excluded ones. This should produce an improved approximate solution (better in terms of volume or mass loaded, depending on the optimization criterion chosen, but with possible intersections) and an expectantly improved subsequent *abstract configuration*. To this purpose, the *packing* module current solution is 'immersed' into a grid domain, giving rise to a set $\hat{N}_G$ of non-covered grid nodes; see Fig. 4.5.

The basic idea of the *hole-filling* module is that of selecting a subset $\hat{N}'_G \subset \hat{N}_G$ and one $\hat{I}'_E \subset \hat{I}_E$, of the currently excluded items, potentially associable to the chosen non-covered nodes (assuming $\left|\hat{I}'_E\right| \leq \left|\hat{N}'_G\right|$). This is aimed at obtaining (even if in an approximate way, i.e. with possible overlap) more loaded volume (or mass). Three sub-steps are then considered hereinafter:

- *Non-covered node selection*
- *Excluded item selection*
- *Overall overlap minimization*

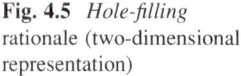

Fig. 4.5 *Hole-filling*
rationale (two-dimensional
representation)

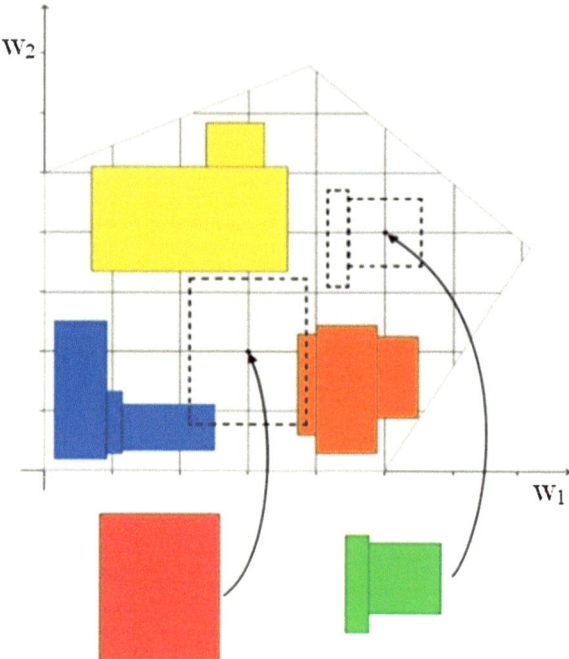

The first and the second are skipped immediately, whenever the relevant sets
have the desired cardinality.

4.3.3.4   Non-covered Node Selection

The following simple MIP model serves the scope of this sub-step. Denoting by ν
the index of the generic non-covered node of $\hat{N}_G$, the binary variable ζ_ν is
introduced, with the meaning $\zeta_\nu = 1$ if the corresponding (non-covered) node is
selected; $\zeta_\nu = 0$ otherwise.

The conditions below are then posed:

$$\sum_{\nu \in \hat{N}_G} \zeta_\nu = |\hat{N}'_G|, \tag{4.5}$$

$$\forall \nu, \nu' \in \hat{N}_G / \nu < \nu' \quad \tilde{\zeta}_{\nu\nu'} \leq \zeta_\nu, \tag{4.6a}$$

$$\forall \nu, \nu' \in \hat{N}_G / \nu < \nu' \quad \tilde{\zeta}_{\nu\nu'} \leq \zeta_{\nu'}, \tag{4.6b}$$

where $\tilde{\zeta}_{\nu\nu'} \in [0, 1]$. The *objective* function below selects the (non-covered) nodes
maximizing the overall relative distance:

$$\max \sum_{\substack{\beta \in B, \\ \nu, \nu' \in \hat{N}_G / \\ \nu < \nu'}} \left(W_{G\beta\nu} - W_{G\beta\nu'} \right)^2 \tilde{\zeta}_{\nu\nu'}. \tag{4.7}$$

Here $W_{G\beta\nu}$ are the coordinates of the grid nodes. It is understood that alternative selection criteria could be chosen. This step provides, as outcome, the subset $\hat{N}_G'$ of the selected non-covered nodes.

4.3.3.5 Excluded Item Selection

In this sub-step, $\hat{N}_{Gi}' \subseteq \hat{N}_G'$ denotes for each item $i \in \hat{I}_E$, the set of selected nodes that allow its placement inside the domain, for at least one orientation $\theta_{\omega i}$. The following binary variables are introduced with the meaning $\xi_{i\nu} = 1$ if the excluded item i is associated to the node ν; $\xi_{i\nu} = 0$ otherwise. The following equations guarantee that at most one item $i \in \hat{I}_E$ is allocated to the same node:

$$\forall i \in \hat{I}_E, \forall \nu \in \hat{N}_G' \sum_{\substack{i \in \hat{I}_E / \\ \nu \in \hat{N}_{Gi}}} \xi_{i\nu} \leq 1. \tag{4.8}$$

The following *objective* function substitutes (2.7) and maximizes the total volume (or mass) of the excluded items, associated to the grid nodes:

$$\max \sum_{\substack{i \in I_E / \\ \nu \in \hat{N}_{Gi}}} K_i \xi_{i\nu}. \tag{4.9}$$

The outcome of this sub-step determines the set of the items to add.

4.3.3.6 Overall Overlap Minimization

To perform this step, either the first or second linear reformulation (Sects. 3.1.1 and 3.1.2) is adopted. For the items already loaded, in the current *packing* module solution, the corresponding *abstract configuration* is imposed. All the *non-intersection* constraints involving the items selected to be added are, instead, generated. In such a way, the *objective* function minimizes the overall overlap. When balancing conditions are present, the relevant constraints are taken into account, together with the following equations:

$$\forall \beta \in B, \forall i \in \hat{I}_E \quad w^*_{\beta i} = \sum_{\nu \in \hat{N}_{Gi}} W_{G\beta\nu}\xi_{i\nu} \qquad (4.10)$$

(where $w^*_{\beta i}$ are the centre of mass coordinates of the new items).

4.3.3.7   General Methodology Background and Alternative Approach

It is understood (see Fig. 4.3) that the *item-exchange* and *hole-filling* modules can be activated in various sequences, following different strategies. The *packing* MIP model is solved, each time, by adopting a *branch-and-bound*. Throughout this process, the binary variables χ, σ and θ are handled sequentially, by groups of items, prioritized by volume (or mass). A *depth-first* strategy is followed, during which subsets of binary variables can temporarily be fixed. A lower bound *cutoff* is set, on the basis of the best-so-far solution, and part of the items, previously picked, can be imposed, following a *greedy* search approach. If a satisfactory solution is found, it is taken as the ultimate one and the whole process ends. Otherwise, the best-so-far solution is stored and the process continues, until a (previously stated) stopping rule intervenes.

It should be further observed that the heuristic process discussed in this section, essentially, reproduces an overall (*delayed*) *column generation* philosophy (consult, e.g. the topical entry of INFORMS Computing Society 2013). The *packing* MIP mode, indeed, at each step, contemplates only a limited subset of variables σ, corresponding to the current *abstract configuration* imposed. This model thus plays the *master's* role, in a *column generation* framework. The generation of 'good' *abstract configurations*, and the corresponding selection of the variables σ, instead, represents the *pricing* phase, carried out by a heuristic process.

Before concluding this section, it is worth noticing that quite a promising alternative, compliant with the overall methodology adopted, consists of utilizing the non-restrictive reformulation of Sect. 3.1.3. This can substitute, tout court, both the *item-exchange* and *hole-filling* modules.

As far as the first is concerned, a number of items are still selected as candidates for the exchanges, as discussed above. Afterwards, all the *relative position* constraints, corresponding to the current *abstract configuration*, with at least one candidate, are substituted with the full set of *non-intersection* inequalities (2.6), (3.5a), (3.5b) and (3.9). The MIP model of the non-restrictive reformulation is hence adopted, keeping, for all the remaining items, the imposed *relative position* constraints. The objective function hence aims at carrying out the advantageous exchanges, improving, if successful, the current solution. In such a case, a new *abstract configuration* is directly generated. This is in general a larger perturbation of the current one, with respect to that obtainable by the *item-exchange* module.

As it is easily gathered, similar considerations hold when the *hole-filling* module is considered. In this case, all the *relative position* constraints corresponding to the items already loaded are kept. Those relevant to the items selected to be supposedly added are, instead, again substituted by the full sets of *non-intersection* constraints. The MIP

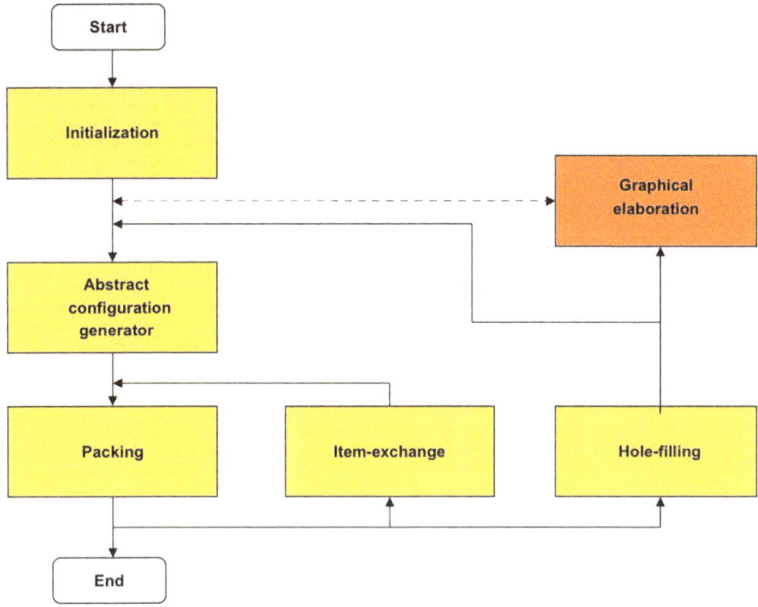

Fig. 4.6 Graphic-numerical interaction

model of the non-restrictive formulation tries and exploits (in a continuous mode) the volume still available and no discretization of the domain is needed any longer. If an improved solution is thus obtained, an upgraded *abstract configuration* is available.

4.3.4   Interaction with the Solution Process

As shown in Sect. 4.2, *abstract configurations* can easily be extracted by approximate solutions of the original problem. These are provided by the *initialization* and *hole-filling* modules. Any alternative process, however, up to yielding 'good' approximate solutions, could be activated as well. If, for example, a solution, of an instance similar to the specific one to sort out, is available, then it could be utilized to generate a first *abstract configuration*, skipping, directly, the *initialization* step. And similar opportunities could come up during the whole solution process, whenever a new *abstract configuration* has to be generated.

This suggests a sort of parallelization of human and computational capabilities by means of a two-way interface between the optimizer and a graphic system. The relevant approach can become quite effective in practice. Intermediate outcomes can, indeed, be visualized, time after time, and the natural human skills, up to managing (even very tricky) three-dimensional jobs by actually 'seeing' the objects involved, profitably exploited. This cannot only speed up the whole search for a satisfactory solution, but it also allows the extemporaneous introduction of further conditions hard to formulate explicitly in the model (for instance, ergonomic conditions). Figure 4.6 illustrates the two-way interface rationale.

Chapter 5
Computational Experience and Real-World Context

Dealing with non-standard packing problems, and precisely because they are by definition outside the framework of any conventional classification, poses, from the experimental point of view, non-negligible difficulties. Firstly, a remarkable effort is requested to collect, or even generate ex novo, non-trivial instances. They need, actually, to cover an adequate number of scenarios, representative of a sufficiently wide area of real-world applications. Secondly, the elaboration of instances of practical interest is mostly very time consuming. Therefore, an extensive dedicated test campaign is extremely demanding, both in terms of human and computational resources.

In this chapter, an attempt has been made to provide useful insights on the computational aspects, relevant to the various formulations and approaches proposed so far. The author is, nonetheless, aware that a systematic and exhaustive experimental approach could hardly be followed. Outcomes concerning the relevant (ongoing) trial activity are reported hereinafter. The case studies are grouped in separate sections, based on different perspectives, also with the expectation of stimulating possible directions for further dedicated research. In the whole chapter, if not otherwise specified, IBM ILOG Optimizer 12.3 (IBM Corporation 2010) is referred to as the MIP solver adopted, supported by a personal computer, equipped with Core 2 Duo P8600, 2.40 GHz processor; 1.93 GB RAM; and MS Windows XP Professional, Service Pack 2.

5.1 Direct Solutions Obtained from the General MIP Model

The heuristic approaches proposed in Chap. 4 have been introduced to obtain satisfactory (suboptimal) solutions to real-world model instances whilst reducing the computational effort as much as possible. As it is easily gathered, most practical

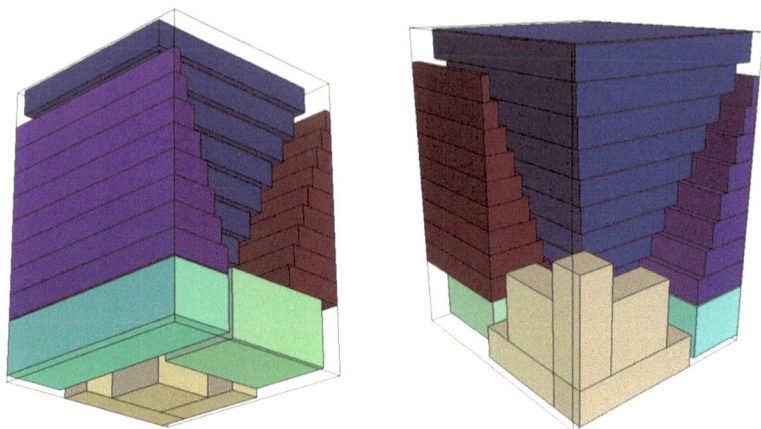

Fig. 5.1 Tetris-like items inside a parallelepiped (Case Study 1.14)

instances can hardly be solved directly, as a matter of fact. This is because of the general MIP model intrinsic difficulties (Sect. 4.1) that are significantly increased when additional conditions (Sect. 2.3) are present. Nonetheless, when relatively small-scale exercises are involved, good-quality results can be obtained as well, by solving the model directly. A set of pertinent case studies (1.1–1.20, see Appendix) has been considered, as a general indication. The packing instances in question are expressed in terms of *feasibility* (i.e., all items have to be loaded). They have been deliberately 'fabricated', in order to deal with cases known a priori for admitting at least one solution.

These case studies contain both single parallelepipeds and actual tetris-like items. All are allowed any possible rotation. The domain is always a parallelepiped, except for Case Study 1.18, for which it is a right prism. The general MIP model of Sect. 2.1 (including some of the auxiliary constraints discussed in Sect. 2.3) has been utilized with a different *objective* function, aimed at minimizing the centre of mass off-centring. For this purpose, the minimum 'virtual' cube, acting as centre of mass domain and 'centred' with respect to the container, is searched for (in Case Studies 1.1–1.4 and 1.7–1.11, the 'virtual' cube has additionally been provided with upper bounds). It is assumed that all items involved are homogeneous and have the same density.

Some details are reported in Tables A.1 and A.2; see Appendix. Figures 5.1, 5.2 and 5.3 represent Case Studies 1.14, 1.18 and 1.20, respectively. As usually understood throughout the whole text, all *components* of each tetris-like item are represented with the same colour.

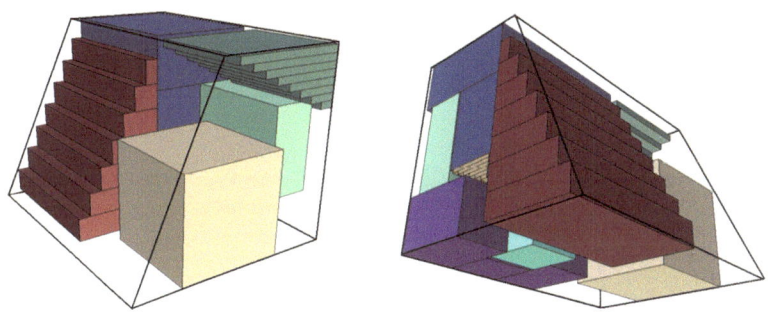

Fig. 5.2 Tetris-like items inside a right prism (Case Study 1.18)

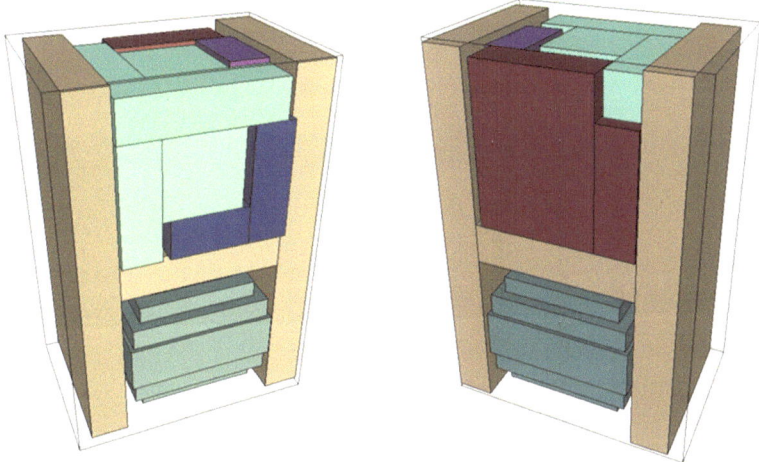

Fig. 5.3 Large tetris-like item acting as a domain (Case Study 1.20)

5.2 Direct Solutions Obtained by Reformulations of the General MIP Model

A number of case studies are considered in this section, focusing on the general MIP model reformulations 3.1.2 and 3.1.3, discussed in Chap. 3. All cases reported in this section focus on the packing of single parallelepipeds, with any possible rotation, into a parallelepiped. No additional conditions have been included (apart from the presence of *separation* planes in Case Study 2.2). All the case studies considered in this section have been solved by utilizing IBM ILOG CPLEX 12.5.1 (supported by a personal computer, equipped with Core 2 Duo P8600, 2.40 GHz processor; 1.93 GB RAM; MS Windows XP Professional, Service Pack 2; and CPLEX 12.5.1 version significantly outperforms 12.3, also referred to in this section).

Table 5.1 Six classes
of test parallelepipeds

Classes of single parallelepipeds	L1 side (units)	L2 side (units)	L3 side (units)
A	4	4	4
B	2	3	5
C	1	3	6
D	1	2	6
E	1	3	3
F	1	2	2

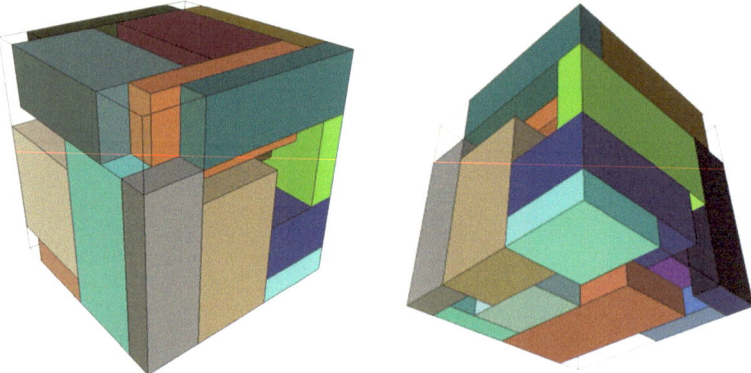

Fig. 5.4 Case Study 2.1

Some interesting case studies, relative to the second linear reformulation for solving the *feasibility* subproblem (Sect. 3.1.2), are reported hereinafter.

Table 5.1 reports six classes of parallelepipeds adopted to execute some tests considered.[1] Case Study 2.1 Contemplates 22 items, extracted from the table: 1 of type A, 6 of B, 6 of C, 5 of D and 4 of E. The domain consists of a cube of eight units. The solution depicted in Fig. 5.4 was found in 30 CPU seconds. The occupied volume reaches 87.5 % of the total available.

Cases Studies 2.2, 2.3 and 2.4 follow. Case 2.2 includes two *separation* planes. Tables A.3, A.4 and A.5, in the Appendix, report, for each case, the item dimensions, whilst Table A.6 those of the relevant domains. The results obtained are summarized here below in Table 5.2. Case Study 2.4 is represented graphically in Fig. 5.5.

The non-restrictive reformulation of the general MIP model reported in Sect. 3.1.3 seems quite suitable to solve the problem directly, when not too large-scale

[1] These classes refer to quite a difficult instance proposed by Jürgen Rietz (Dept. Produção e Sistemas, Centro de Investigação Algoritmi da Universidade do Minho, Escola de Engenharia, Universidade do Minho, 4710–057, Braga, Portugal). It consists of the following: given a cube of 8 units, load 1 item of type A and 6 for all of the remaining types.

Table 5.2 Results of Case Studies 2.2 to 2.4

Case studies	Total number of single parallelepipeds	Loaded volume % (rounded to nearest)	CPU time (s)
Case Study 2.2	31	80.9	2,659
Case Study 2.3	25	84.6	549
Case Study 2.4	17	90.5	834

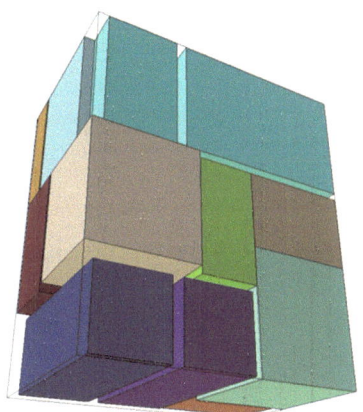 

Fig. 5.5 Case Study 2.4

instances are involved. As is easily seen, this reformulated model, differently from the heuristic approaches considered in Chap. 4, allows the optimizer, at least theoretically, to find and to prove the actual optimal solution.

An experimental analysis in this direction is currently ongoing. A significant effort is still expected to confirm the apparent advantage of the use of this reformulation, both as a stand-alone model and in support of the heuristic approaches.

Some successful exercises are reported here. A first indication can be provided by reconsidering Case Studies 2.2–2.4, as solved by the non-restrictive reformulation in question. The same results were obtained in terms of volume occupation, packing all the given items, even if no impositions were made on their loading. Different outcomes, nonetheless, arose, concerning the computational effort: Case Study 2.2 was solved in 280 CPU seconds, Case Study 2.3 in 370 CPU seconds and Case Study 2.4 in 313 CPU seconds (the information available to date, relevant to the non-restrictive reformulation, is however not sufficient to confirm this apparently outperforming trend).

Fig. 5.6 Case Study 3.1

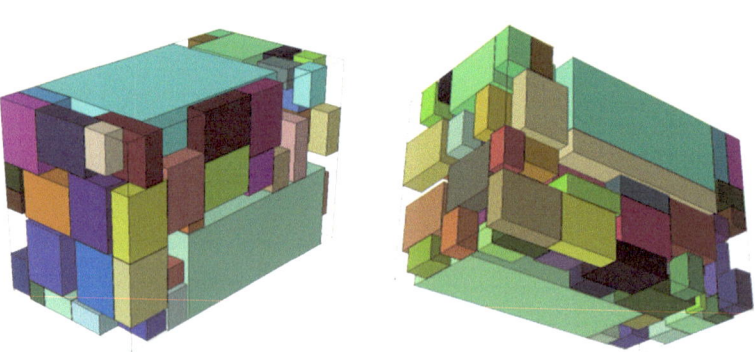

Fig. 5.7 Case Study 3.3

Case Study 3.1 refers to the instance derived from Table 5.1, involving 1 item of class A, 6 of B, 6 of C, 6 of D, 6 of E and 6 of F, considering, as a domain, a cube of 8 units (cf. Note 2). An optimal solution (loading all the given 31 items) was found in 995 CPU seconds. The occupied volume is 98 % of the available one. It is depicted in Fig. 5.6.

Input data relevant to the following Case Studies 3.2 and 3.3 are reported in Tables A.7 and A.8; see Appendix. An optimal solution, including all the 51 items, was found in 757 CPU seconds for Case Study 3.2. The occupied volume is 82.7 % of the available one. An optimal solution, including all the 84 items, was found in 2,727 CPU seconds for Case Study 3.3. The occupied volume is 80 % of the available one. It is worth noticing that the relative instance contains 27,902 constraints and 22,261 variables, of which 21,756 are binary. It is represented by Fig. 5.7.

5.3   Use of the Linear Reformulations to Obtain Approximate Solutions

In the following, some insights are provided, concerning the (*LP-relaxed*) reformulations of Sects. 4.3.1.1 and 4.3.1.2, respectively, both aimed at finding approximate solutions (as initialization steps). Case Studies 2.2–2.4 (cf. Sect. 5.2) are reconsidered here as reference instances.

The first linear reformulation has been utilized to find an approximate solution to Case Study 2.3. The process took only 2 CPU seconds, but 38 intersections were identified (out of 300 pairs of items). The overlap volume is 38.6 % of that associated to the totality of items to load. The graphical results are represented by Fig. 5.8.

The second linear reformulation was adopted by dropping inequalities (3.6) (of Sect. 3.1.2) and including (4.2) (as suggested in Sect. 4.3.1.2). An approximate solution to Case Study 2.4 was obtained within a time limit of 300 CPU seconds, with 23 intersections (out of 136 pairs of items) and 12.7 % of overall volume overlap, considering the actual parallelepipeds. It is shown in Fig. 5.9 (with the occurring intersections). The variation of the second linear reformulation of Sect. 3.1.2 (substituting (3.6) with (3.8), cf. Sect. 4.3.1.2) was, instead, considered for Case Study 2.2. The relative solution (suboptimal for this reformulated model) is represented in Fig. 5.10 (with the occurring intersections). It was found within a time limit of 300 CPU seconds. The number of identified intersections (with respect to the actual items) is 11 (out of 528 pairs of items, considering also the *separation* planes as such), corresponding to 5.5 % of the overlap volume.

The three study cases in question suggest that the first linear reformulation provides quick but quite imprecise solutions, whilst the two versions of the second reformulation are more time consuming but offer better results. This trend seems to be confirmed by the comparative analysis currently ongoing, but a further in-depth experimentation is certainly needed. The approach adopted for Case Study 2.2 presents, for the time being, promising perspectives. From the preliminary outcomes available, indeed, it is usually able to find a number of *integer-feasible* (suboptimal) solutions quite easily.

Fig. 5.8 Case Study 2.3 approximate solution (obtained by the first linear reformulation)

Fig. 5.9 Case Study 2.4 approximate solution (obtained by the second linear reformulation)

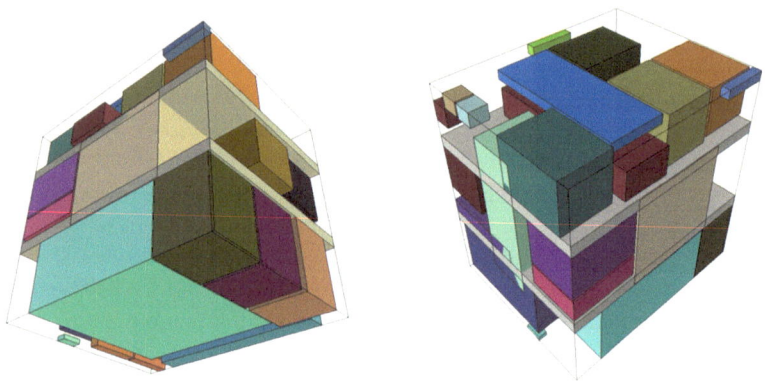

Fig. 5.10 Case Study 2.2 approximate solution (obtained by the second linear reformulation variation)

5.4   Nonlinear Reformulation Approach to Improve Approximate Solutions

A dedicated experimental analysis, focusing on the use of the nonlinear reformulation referred to in Sect. 4.3.1.3, with *objective* function (3.13), is currently ongoing (Fasano and Castellazzo 2013). As outlined there, this approach can be utilized to improve the approximate solutions, obtained either by the first or second linear reformulations; cf. Sects. 4.3.1.1 and 4.3.1.2, respectively (or by any alternative initialization process).

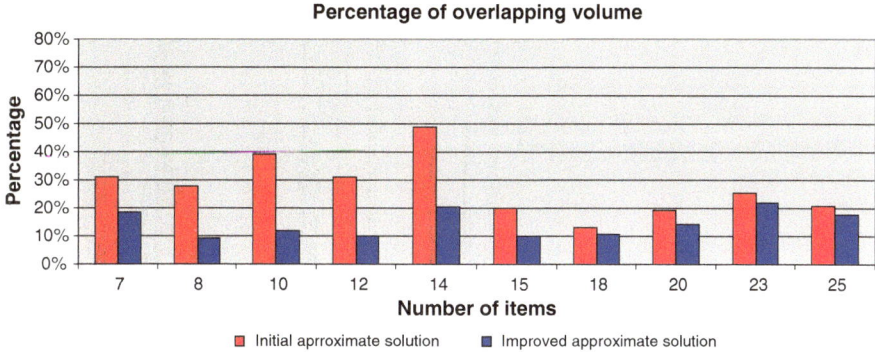

Fig. 5.11 Initial overlap volume reduction

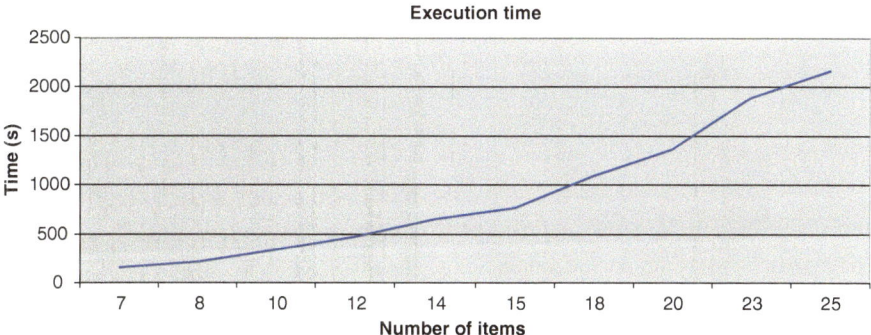

Fig. 5.12 Computational effort

Some preliminary outcomes are briefly reported here. A first rough trend estimate, relative to a sample of 35 case studies (considering, for the time being, just quite a limited number of single parallelepipeds), is suggested by Fig. 5.11 that shows different groupings, based on the number of items involved. For each group, the (average) percentage of overlap volume is displayed, with respect to the initial approximate solution (left column) and the improved one (right column). Some indications, concerning the computational effort, are given in Fig. 5.12. Such an effort is expected to decrease in the near future, by improving the optimization strategies adopted.

Three case studies (4.1–4.3) involving 7, 12 and 18 items are depicted by Figs. 5.13, 5.14 and 5.15, respectively (representing the initial solutions on the left and the improved ones on the right).

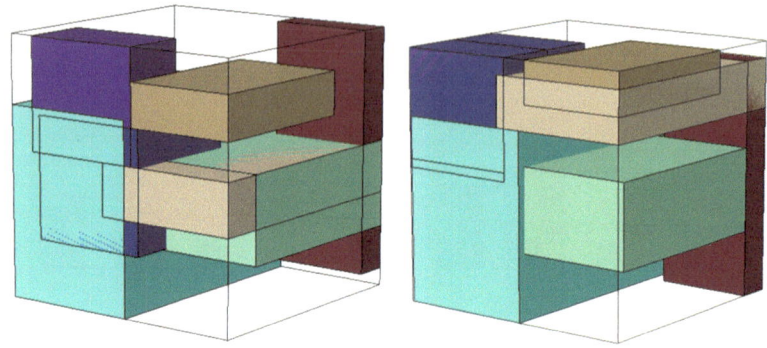

Fig. 5.13 Case Study (4.1) with 7 items

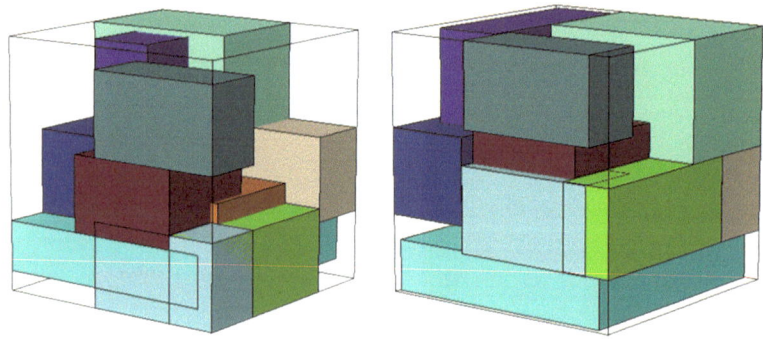

Fig. 5.14 Case Study (4.2) with 12 items

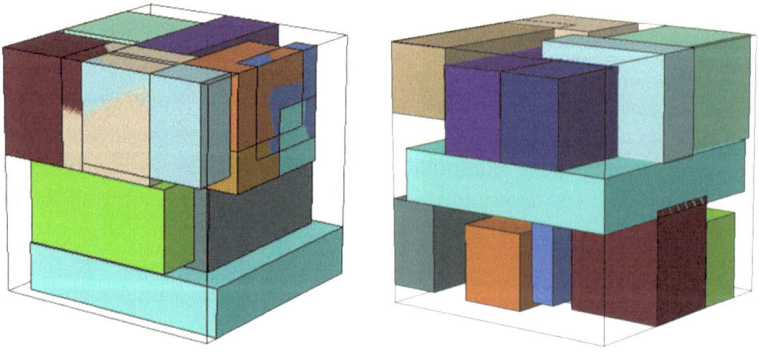

Fig. 5.15 Case Study (4.3) with 18 items

5.5   The Use of Heuristics

This section is devoted to providing some insights on the use of the heuristic approaches proposed in Chap. 4. A significant number of real-world packing issues (more or less complicated, in terms of additional conditions) have been solved successfully in the space engineering context that gave rise to this work (in the framework of the International Space Station, ISS, cf. http://www.nasa.gov, in particular within the CAST project; see (Fasano et al. 2009)). Nonetheless, a substantial commitment is still expected to consolidate the outcomes available to date.

As previously pointed out, differently from other methods, the modeling-based ones, proposed in this volume, try to solve the relevant packing models, taking into account, contemporarily, all the items (or at least subsets of the original instance). This offers the evident advantage of providing a global point of view, allowing, if necessary, for the introduction of overall (i.e. 'transversal') conditions, such as balancing. On the other side, it is easily seen that solving such (MIP/MINLP) models is generally much more complex than carrying out packing algorithms based on sequential placement. It is hence quite obvious that if no additional conditions have to be taken into account, then most non-modeling-based approaches (e.g., Martello et al. 2000) are expected to outperform the modeling-based ones discussed in this monograph.

In addition to the above considerations, a non-trivial issue comes up, since a practical threshold, concerning the scale of the instances to face, is understood. The number of items involved becomes, as a matter of fact, a first limiting factor. It, indeed, directly affects the instance size (that, when the model is formulated in terms of MIP, corresponds to that of the relative matrix), as well as the number of binary variables. Nevertheless, even when the given instance is, as a matter of fact, too large to cope with, a partition into subproblems can be carried out. Quite a successful approach of this type has been applied, in the space engineering context, to the Automated Transfer Vehicle (ATV, ESA; cf. http://www.esa.int). Its extremely challenging cargo accommodation problem has been tackled by developing an ad hoc packing optimization system, decomposing the overall problem at different levels (see Fasano et al. 2009).

5.5.1   Test Instances from the Literature

The description of instances containing a significant number of *tetris*-like items, with several *components* each, represents quite a heavy task indeed. For this reason and in order to provide the reader with quite an easy-to-access experimental framework, it has been decided to concentrate on standard instances.

A dedicated test campaign has been fulfilled for the first heuristic procedure presented in Sect. 4.3.2 (at present, the most consolidated approach, from the

Table 5.3 Results of 670 case studies (heuristic procedure of Sect. 4.3.2)

	Set 1	Set 2	Set 3	Set 4	Set 5	Set 6	Set 7
Average volume exploitation (%)	78.13	78.48	79.60	79.75	80.04	80.05	80.01
Worst case volume exploitation (%)	69.13	68.94	71.64	72.25	72.56	71.17	72.60
Best case volume exploitation (%)	86.68	85.41	86.79	86.67	86.67	86.61	86.99
Average loaded items	83	80	79	78	80	78	77
Worst case loaded items	42	45	50	47	53	52	56
Best case loaded items	138	122	123	130	118	107	105

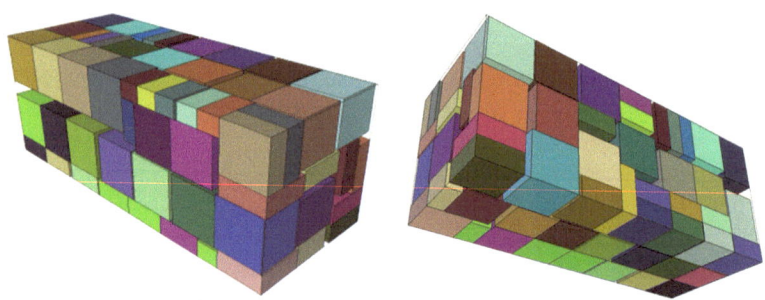

Fig. 5.16 Case Study 5.1.43

experimental point of view), referring to the 'Three Dimensional Cutting and Packing Data Sets—THPACK 1–7 BR' (Bischoff and Ratcliff 1995): http://www.euro-online. org/web/ewg/25/esicup-euro-special-interest-group-on-cutting-and-packing. As it is known, this test bed consists of 7 sets of 100 instances each. They are indicated in the following as Case Studies '5.s.n', where 's' is the progressive number of the set and 'n' the index of the test problem instance. 670 instances out of the 700 available have been tested (excluding all those exceeding 200 items).

Table 5.3 reports the relevant results. All the tests were executed within the limit of 1 h of CPU time. Figure 5.16 shows as an indicative example, Case Study 5.1.43 solution, in which 86.68 % of the available volume has been exploited (loading 94 items out of 141).

The experimental activity relevant to the heuristic procedure described in Sect. 4.3.3 is currently being carried on. Since the computational effort strongly depends on the overall solution strategy followed, it is worth providing some insights on the relevant performances, at each single-phase level. On the basis of the experience acquired to date, Table 5.4 provides quite a consolidated general trend, referring to the process steps separately.

As far as the alternative approach suggested in Sect. 4.3.3 is concerned, a further set of 24 tests from 'Three Dimensional Cutting and Packing Data Sets—THPACK 1–7 BR' was executed, namely, 5.1.17, 5.1.39, 5.1.67, 5.1.68, 5.1.76, 5.1.91, 5.1.100; 5.2.4, 5.2.13, 5.2.39, 5.2.59, 5.2.77, 5.2.79, 5.2.85, 5.2.96; 5.3.39, 5.3.56, 5.3.59, 5.3.77; 5.4.39, 5.4.56, 5.4.79; 5.5.56; 5.6.13.

Table 5.4 Computational trend at single-step level (heuristic procedure of Sect. 4.3.3)

Steps	Involved items	CPU time estimates (s)
Initialization	75–100	45–90 (recursive mode)
Abstract configuration generation	75–100	<5
Packing	75–100	30–60
Hole-filling	10–15	<15
Item exchange	10–15	<5

Table 5.5 Results of 24 case studies (heuristic procedure of Sect. 4.3.3)

Case studies	Total no of items (parallelepipeds)	No of loaded items exploiting 75 % of the available volume	CPU time to reach the 75 % of the available volume
5.1.17	213	116	00:16:36
5.1.39	243	166	00:13:23
5.1.91	238	88	00:07:30
5.1.100	214	142	00:32:41
5.2.4	201	94	00:08:05
5.2.13	228	109	00:22:10
5.2.79	206	128	00:49:53
5.2.85	209	121	00:22:50
5.2.96	202	120	00:31:09
5.3.56	212	112	00:13:49
5.3.77	201	131	01:03:34
5.4.56	233	130	00:31:49
5.4.79	217	119	00:57:58

For all these tests the number of items available is within the range between 201 and 275. To solve this large-scale instance successfully, an ad hoc solution strategy was thought up. A basic cycle, consisting of the following sequence of steps was introduced: *initialization, packing, item-exchange* and *hole-filling*. A number of basic cycles were allowed to be executed, by extending incrementally the set of items involved, until 75 % of the occupied volume was reached. Afterwards, a final cycle consisting of *hole-filling* steps only was admitted. Within every cycle, each of the above steps was allowed to be repeatedly fulfilled, following appropriate stopping rules. The process had (about) 1 CPU hour as a time limit. Overall results are reported here below:

- Average volume = 76.31 %
- Worst case volume exploitation = 61.19 %
- Best case volume exploitation = 85.41 %
- Average loaded items = 141
- Worst case loaded items = 103
- Best case loaded items = 182

Table 5.5 shows the CPU time requested to attain (when reached) 75 % of volume exploitation and the corresponding number of items loaded.

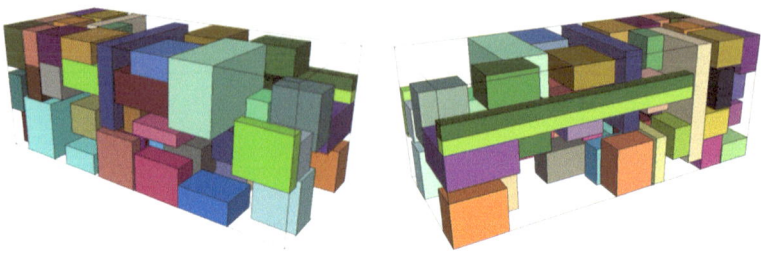

Fig. 5.17 Case Study 6—third cycle solution (50 % of exploited volume)

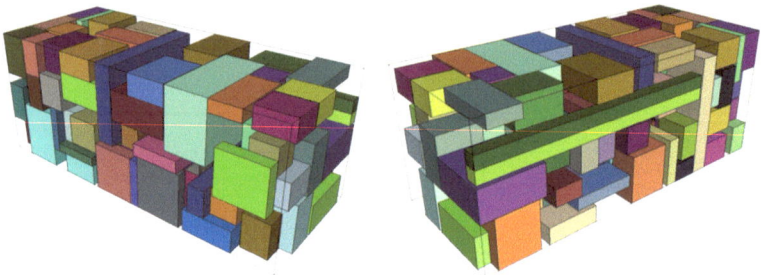

Fig. 5.18 Case Study 6—fifth cycle solution (65 % of exploited volume)

Fig. 5.19 Case Study 6—solution obtained in 1 CPU hour (85.77 % of exploited volume)

5.5.2   *Instance Adopted to Tune the Solution Strategy*

A 'fabricated' instance (Case Study 6) was introduced (in addition to others) to tune
the solution strategy outlined above. Representing an interesting exercise solved
successfully, it is briefly considered hereinafter. This instance consists of the
packing of up to 280 parallelepipeds into a parallelepiped of dimensions 540, 225
and 220 units, respectively, maximizing the loaded volume. Results obtained, with
1 CPU hour as a time limit, throughout the whole procedure are illustrated in
Figs. 5.17, 5.18 and 5.19, showing different solution levels. Relevant details are
reported in Table A.10; see Appendix.

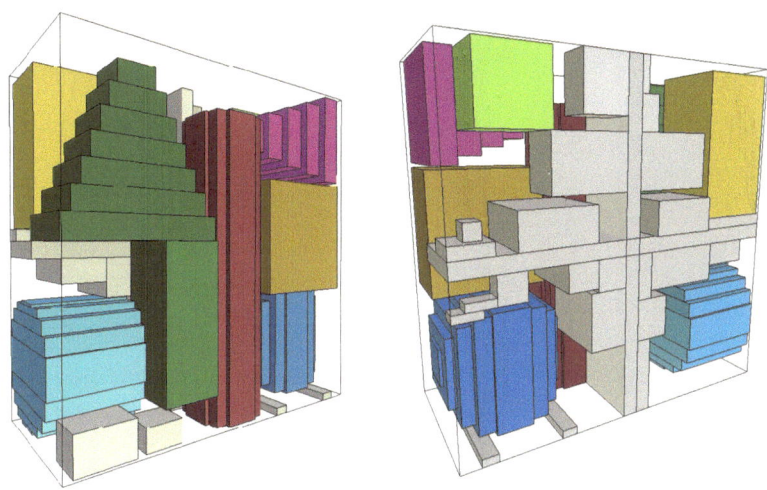

Fig. 5.20 Case Study 7.1 (with *structural* elements and forbidden zones)

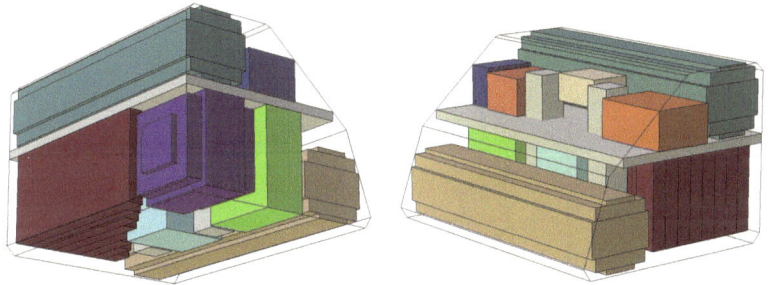

Fig. 5.21 Case Study 7.2 (with a curved domain, a separation plane and *structural* elements)

5.5.3 Close-to-Real-World Instances

Before concluding this section, two similar-to-real-world instances (Case Studies 7.1 and 7.2) successfully solved, in support of the ISS and ATV logistics, are shown in the following; see Figs. 5.20 and 5.21 (the relevant technical details are kept confidential). They were solved by utilizing the heuristic procedure of Sect. 4.3.2, requiring about 500 CPU seconds.

Further applications involving *tetris*-like items are considered in Sect. 6.1.3.

Chapter 6
Extensions and Mixed-Integer Nonlinear Approaches for Further Applications

The modelling approach advocated by this volume is susceptible to possible extensions. One, in particular, deals with the problem of looking into how the free volume of a container, partially loaded with *tetris*-like items, could be profitably exploited. *Virtual* items, i.e. (rectangular) parallelepipeds of non-prefixed dimensions, are purposely introduced to 'suggest' how to fill the empty volumes. In order to meet the practical demand, depending on the specific framework in question, limitations are stated on the maximum number of *virtual* items allowed, as well as on their minimum dimension. This issue is investigated hereinafter, highlighting a dedicated MIP formulation (Sect. 6.1).

The *global optimization* approach, stressed in this work, is further extended to the problem of packing simple polygons, with continuous rotations, inside a convex polygon. A heuristic approach, solving recursively a dedicated *mixed-integer nonlinear programming* (MINLP) model (founded on necessary conditions), is outlined (Sect. 6.2). It is aimed at providing an approximate global solution that can be further refined by exact local optimization-based methods. The *tetris*-like formulation is properly adapted to generate the first starting solution and profitably initialize the *mixed integer nonlinear programming* search process.

6.1 Exploiting Empty Volumes by Adding Virtual Items

This section is devoted to the issue of exploiting the residual volume of a container, partially loaded, by adding a certain number of *virtual* items. These are intended as (rectangular) parallelepipeds, not defined a priori in terms of dimensions. They are aimed at indicating how real items could still be loaded into the container.

This kind of problem arises, for instance, quite frequently in the framework of the logistic support to the International Space Station (ISS, cf. http://www.nasa. gov), when planning the periodical resupply of the resources stowed on board. A significant number of similar applications are expected in logistics in general. Here, indeed, the frequent necessity of introducing (rigid) packaging material, to

G. Fasano, *Solving Non-standard Packing Problems by Global Optimization and Heuristics*, SpringerBriefs in Optimization, DOI 10.1007/978-3-319-05005-8_6, © Giorgio Fasano 2014

prevent item collisions, represents, from the analytical point of view, the same problem typology. Further examples, albeit in quite different fields, concern the use of autonomous robots (e.g. when requested to determine accessibility zones or to carry out assembling activities) and specific packing issues in the VLSI context.

The problem considered in this section can be formulated as follows:

Given a (convex three-dimensional) domain D, and a set of tetris-like items inside it, let us add a number (not exceeding the maximum value of $\overline{N}_V$) of virtual items (i.e. rectangular parallelepipeds of variable dimensions). The total loaded volume is maximized by repositioning, if necessary, the (tetris-like) items already accommodated.

Two classes of items are then taken into account: the tetris-like and *virtual*, that are single (rectangular) parallelepipeds, with no a-priori-given dimensions. All positioning rules of the general problem of Sect. 2.1, i.e. *orthogonality*, *domain* and *non-intersection* conditions, still hold. In particular:

- *Each virtual item has to be positioned orthogonally, with respect to the main reference frame.*
- *Each virtual item has to be contained within (the convex domain) D.*
- *Virtual items cannot overlap either with the tetris-like or other virtual ones.*

6.1.1   Model Formulation

An MINLP formulation of the optimization problem in question is considered first. We shall point out that the packing rules expressed above can be grouped as follows:

- *Orthogonality*, *domain* and *non-intersection* conditions for tetris-like items only
- *Orthogonality*, *domain* and *non-intersection* conditions for *virtual* items only
- *Non-intersection* conditions between tetris-like items and the *virtual* ones

As far as the first group is concerned, they are represented by constraints (2.1), (2.2), (2.3), (2.4), (2.5a), (2.5b) and (2.6) of Sect. 2.1. The *orthogonality*, *domain* and *non-intersection* conditions for *virtual* items only, as well as the *non-intersection* ones, between tetris-like items and the *virtual* ones, are quite straightforward. They are discussed here below.

Let us introduce the set of *virtual* items I_V, and the binary variables $\chi_{Vj} \in \{0, 1\}$, $j \in I_V$, with the meaning: $\chi_{Vj} = 1$ if *virtual* item j is included; $\chi_{Vj} = 0$ otherwise. For each *virtual* item j, $w_{V0\beta j}$ denote the centre coordinates, with respect to the main reference frame, and $l_{V\beta j}$ the side parallel to the axis w_β (of the main reference frame). E_{Vj} is the set of (eight) vertices associated to j whose coordinates are expressed as follows:

$$\forall \beta \in B, \forall j \in I_V, \forall \eta \in E_{Vj}$$

$$w_{V\beta\eta j} = w_{V\beta0 j} \mp \frac{1}{2} l_{V\beta j}. \tag{6.1}$$

It is easily seen (cf. Sect. 2.1) that the *orthogonality* conditions are implicitly contemplated by equations (6.1), while the *domain* constraints have the same expressions of equations (2.3) and (2.4) (adopting the specific *virtual* item symbolism).

The following inequalities represent the *non-intersection* conditions between the generic tetris-like item i and the *virtual* one j, cf. constraints (2.5a), (2.5b) and (2.6):

$$\forall \beta \in B, \forall i \in I, \forall j \in I_V, \forall h \in C_i$$

$$w_{\beta0hi} - w_{V\beta0j} \geq \frac{1}{2} \sum_{\omega \in \Omega} \left(L_{\omega\beta hi} \vartheta_{\omega i} \right) + \frac{1}{2} l_{V\beta j} - D_{\beta} \left(1 - \sigma^+_{V\beta hij} \right), \tag{6.2a}$$

$$\forall \beta \in B, \forall i \in I, \forall j \in I_V, \forall h \in C_i$$

$$w_{V\beta0j} - w_{\beta0hi} \geq \frac{1}{2} \sum_{\omega \in \Omega} \left(L_{\omega\beta hi} \vartheta_{\omega i} \right) + \frac{1}{2} l_{V\beta j} - D_{\beta} \left(1 - \sigma^-_{V\beta hij} \right), \tag{6.2b}$$

$$\forall i \in I, \forall j \in I_V, \forall h \in C_i \quad \sum_{\beta \in B} \left(\sigma^+_{V\beta hij} + \sigma^-_{V\beta hij} \right) \geq \chi_i + \chi_{Vj} - 1, \tag{6.3}$$

$$\forall i \in I, \forall j \in I_V, \forall h \in C_i \quad \sum_{\beta \in B} \left(\sigma^+_{V\beta hij} + \sigma^-_{V\beta hij} \right) \leq \chi_i, \tag{6.4a}$$

$$\forall i \in I, \forall j \in I_V, \forall h \in C_i \quad \sum_{\beta \in B} \left(\sigma^+_{V\beta hij} + \sigma^-_{V\beta hij} \right) \leq \chi_{Vj}, \tag{6.4b}$$

where $\sigma^+_{V\beta hij}$ and $\sigma^-_{V\beta hij} \in \{0, 1\}$. The *non-intersection* constraints for *virtual* items only are immediately understood. The lower bound $\underline{L}_V$ is further introduced for all *virtual* item sides, in order to obtain acceptable solutions from a practical point of view (i.e. excluding 'too small' objects). The following constraints are thus stated:

$$\forall j \in I_V \quad \underline{L}_V \chi_{Vj} \leq l_{V\beta j} \leq D_{\beta} \chi_{Vj}. \tag{6.5}$$

Since the total volume of the *virtual* items added has to be maximized, the nonlinear *objective* function below is defined:

$$\max \sum_{j \in I_V} \prod_{\beta \in B} l_{V\beta j}. \tag{6.6}$$

Remark 6.1 It is gathered that the additional conditions discussed in Sect. 2.3 can easily be introduced (with proper adaptation, if necessary). As far as the static balancing ones (Sect. 2.3.4.1), in particular, are concerned, they have to be adequately extended to include the virtual items. To do this, we shall firstly assume that they have the (hypothetical average) density R_V. Expressions (2.23) and (2.24) are therefore modified as

$$\forall \beta \in B \quad \sum_{i \in I} M_i w^*_{\beta i} + R_V \sum_{j \in I_V} w_{V\beta 0 j} \prod_{\beta \in B} l_{V\beta j} = \sum_{\gamma \in V^*} V^*_{\gamma \beta} \psi^*_\gamma, \quad \sum_{\gamma \in V^*} \psi^*_\gamma = m,$$

where $m = \sum_{i \in I} M_i + R_V \sum_{j \in I_V} \prod_{\beta \in B} l_{V\beta j}, \ \forall \ \gamma \in V^*, \ \psi^*_\gamma = \widetilde{\psi}^*_\gamma m$ and $\widetilde{\psi}^*_\gamma \geq 0$. As is easily seen, these conditions, differently from the case of Sect. 2.3.4.1, are no longer linear.

6.1.2   Model Approximations

A possible (quite daring) approximation of the MINLP model presented in Sect. 6.1.1 consists of adopting the following linear *objective* function as a *surrogate* of the nonlinear (6.6):

$$\max \sum_{\substack{\beta \in B/ \\ j \in I_V}} l_{V\beta j}. \tag{6.7}$$

An alternative approach consists of replacing function (6.6) with a separable one and carrying out a piecewise linear approximation of each term (e.g. Williams 1993). This can easily be achieved by introducing the (likewise) *surrogate objective* function:

$$\max \sum_{\substack{\beta \in B, \\ j \in I_V/ \\ l_{V\beta j} > 0}} \ln\left(l_{V\beta j}\right). \tag{6.8}$$

This is indeed separable (and no longer a *surrogate* one, when just a single *virtual* item is considered, cf. Sect. 6.1.3). The piecewise linear approximation of each (single-variable) logarithmic term in (6.8) reduces then the original MINLP model to a much simpler (approximate) MIP one. A straightforward formulation is outlined here below (cf. Williams 1993 and Sect. 2.3.2).

For each axis w_β, we shall discretize the intervals $\left[\underline{L}_V, D_\beta\right]$ in a set $D_{SV\beta}$ of subintervals $[D_{V\beta(\gamma-1)}, D_{V\beta\gamma}]$ and then pose

$$\forall \beta \in B, \forall j \in I_V \quad l_{V\beta j} = \sum_{\gamma \in D_{SV\beta}} D_{V\beta\gamma} \lambda_{V\beta\gamma j}, \tag{6.9}$$

$$\forall \beta \in B, \forall j \in I_V \quad \ln\left(l_{V\beta j}\right) \approx \sum_{\gamma \in D_{SV\beta}} \ln\left(D_{V\beta\gamma}\right)\lambda_{V\beta\gamma j}, \tag{6.10}$$

$$\forall \beta \in B, \forall j \in I_V \quad \sum_{\gamma \in D_{SV\beta}} \lambda_{V\beta\gamma j} = \chi_{Vj}, \tag{6.11}$$

where the terms $\lambda_{V\beta\gamma j}$ are nonnegative variables.

It is worth noticing that, in this specific case (as in that of equations (2.20) of Sect. 2.3.2), the *adjacency* condition (for which *at most two adjacent* λ *can be non-zero*) may be dropped tout court (cf. Williams 1993), with significant computational benefit. It is, indeed, sufficient to observe that the optimization problem in question is equivalent to that of minimizing a convex *objective* function. This is immediately seen simply considering that expression (6.8) is equivalent to min $\sum_{\substack{\beta \in B, \\ j \in I_V}} \left[-\ln\left(l_{V\beta j}\right)\right]$

that is convex, as it is a sum (with positive coefficients) of convex functions (e.g. Minoux and Vajda 1986).

Both *objective* functions (6.7) and (6.8) are suitable for providing a starting approximated solution for the (exact) MINLP formulation of Sect. 6.1.1. A more refined (even if more demanding) approach could be followed as an alternative to avoid the introduction of *surrogate* functions. It is based on the method of converting products of (two or more) variables into *separable* functions, by means of quadratic terms (e.g. Williams 1993). It is briefly outlined here.

In the case of the product of two variables $q_1 q_2$, it is sufficient to introduce the new variables s_1 and s_2 (not restricted to be nonnegative), by performing the transformations $s_1 = \frac{1}{2}(q_1 + q_2)$, $s_2 = \frac{1}{2}(q_1 - q_2)$. The terms $q_1 q_2$ are hence substituted with $s_1^2 - s_2^2$ (that is a non-convex function). The method can be extended when the products involve more than two variables and a piecewise linear approximation of the quadratic terms can hence be achieved.

Remark 6.2 The task of minimizing the container area/volume can also be achieved by performing a logarithmic transformation and a piecewise linear approximation, without introducing any surrogate objective function (e.g. Pan and Liu 2006; Wang and Tsai 2010). The resulting model, nonetheless, appears quite complicated. Indeed, indicating with d_β the variables representing the container (parallelepiped) dimensions, with respect to the corresponding axes w_β, the objective function (in logarithmic form) $\sum_{\beta \in B} \ln\left(d_\beta\right)$ to minimize is not convex. As a consequence, the adjacency condition cannot be neglected.

Remark 6.3 The presence of the binary variables χ_{Vj} in expressions (6.5) guarantees that if a virtual item is not added, its contribution to the total volume is zero. This implication is nonetheless implicitly stated by equations (6.1) together with the domain constraints for virtual items.

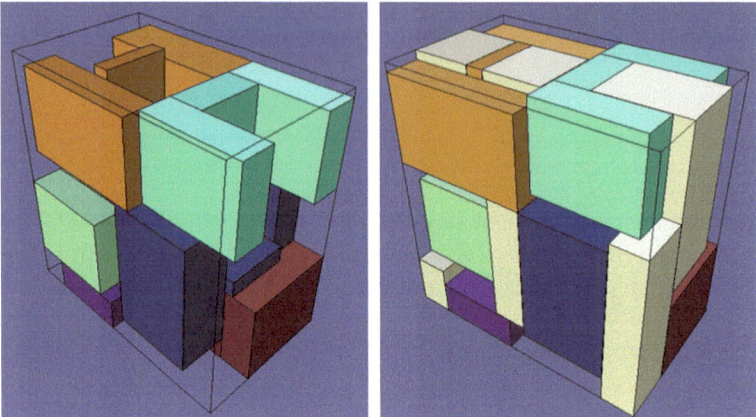

Fig. 6.1 *Virtual* item Case Study 6

6.1.3 Applications

A previous work (Fasano and Vola 2013) focuses on the utilization of the *surrogate* linear *objective* function (6.7), in the context of a dedicated heuristic approach. This has been conceived with the aim of obtaining quick satisfactory (but typically suboptimal) solutions to the original (nonlinear) problem. This approach is outlined, briefly, hereinafter, whilst the reader is referred to the work quoted above for more details, both on the algorithmic and experimental aspects.

An appropriate lower bound, as stated by conditions (6.5), is generated each time, depending on the specific instance to solve. The heuristic approach adopted assumes that an *abstract configuration*, relative to the already loaded tetris-like items, is provided. Since the addition of several *virtual* items, all together, would represent a significant computational effort, the heuristic proce- dure progresses incrementally. This is obtained by adding one *virtual* item after the other, until either a satisfactory solution is obtained or their maximum number is reached.

A currently ongoing experimental analysis (Fasano and Vola 2013) is being carried out. Some insights on the computational results, obtained to date, are briefly illustrated in Table A.11; see Appendix. They refer to a set of 32 case studies. For all tests considered, a maximum threshold of 10 *virtual* items was imposed, setting a runtime limit of 3 CPU hours. Case studies 6 and 22 are illustrated by Figs. 6.1 and 6.2, respectively (on the left the already loaded tetris-like items are shown and, on the right, the *virtual* ones added).

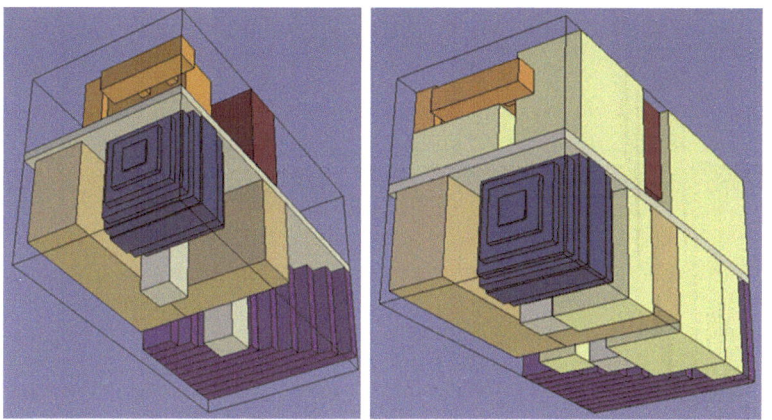

Fig. 6.2 *Virtual* item Case Study 22

6.1.3.1 Container Area/Volume Minimization by Maximizing Virtual Items

The concept of *virtual* item and the relative packing model can be utilized to solve the important issue of the container area/volume minimization. In its two-dimensional form, it consists of placing (orthogonally) a given set of single rectangles into one of minimum area (the generalization to the case of three-dimensional tetris-like items inside a parallelepiped is straightforward). We introduce first the domain D of sides D_β, $\beta = \{1, 2\}$ (the set of axes is still denoted by B, cf. Sect. 2.1), assumed sufficiently big to allow the loading of all items inside the corresponding rectangle. For all of them the constraints stated in the special case of Sect. 2.1 hold, with the corresponding variables χ set to one. Two *virtual* items are then introduced with the purpose of restricting the domain from the right and the upper edges, respectively. They are denoted as $\overline{R}$ and $\overline{U}$. $\overline{R}$ sides are $l_{1\overline{R}}$ (variable), parallel to D_1 and D_2 (constant). Its centre coordinates are indicated with $w_{\beta\overline{R}}$. The following specific (*domain*) constraints are posed:

$$w_{1\overline{R}} + \frac{1}{2} l_{1\overline{R}} = D_1, \tag{6.12a}$$

$$w_{2\overline{R}} = \frac{1}{2} D_2. \tag{6.12b}$$

Similar equations are set for the *virtual* item $\overline{U}$:

$$w_{1\overline{U}} = \frac{1}{2} \left(D_1 - l_{1\overline{R}} \right) \tag{6.13a}$$

$$w_{2\overline{U}} + \frac{1}{2}l_{2\overline{U}} = D_2 \qquad (6.13b)$$

The *non-intersection* constraints (6.2a) and (6.2b) are specified as follows:

$$\forall i \in I \quad w_{1\overline{R}} - w_{1i} \geq \frac{1}{2}\sum_{\alpha \in A}(L_{\alpha i}\delta_{\alpha 1 i}) + \frac{1}{2}l_{1\overline{R}}, \qquad (6.14a)$$

$$\forall i \in I \quad w_{2\overline{U}} - w_{2i} \geq \frac{1}{2}\sum_{\alpha \in A}(L_{\alpha i}\delta_{\alpha 2 i}) + \frac{1}{2}l_{2\overline{U}}, \qquad (6.14b)$$

where the meaning of the symbols is understood (cf. Sect. 2.1, special case). It is immediate to see that, as a consequence, all the given items are confined to the inside of the rectangle cut out from D by $\overline{R}$ and $\overline{U}$. This rectangle has the vertex (0,0) in common with the domain D and the sides $D_1 - l_{1\overline{R}}$ and $D_2 - l_{2\overline{U}}$ laying on w_1 and w_2, respectively. The optimization problem under consideration consists hence of minimizing this area (obviously included within D) simply by adopting, for the *virtual* items $\overline{R}$ and $\overline{U}$, the *objective* function (6.6). This assumes, in the specific case, the following form (easily reducible to a *separable* function, even if non-convex, as outlined in Sect. 6.1.2):

$$\max\left(D_2 l_{1\overline{R}} - l_{1\overline{U}}l_{2\overline{U}}\right). \qquad (6.15)$$

The heuristic approach presented in this section can thus be well adopted to obtain an approximate solution.

6.2   Non-orthogonal Packing of Non-rectangular Items

The key idea advocated in this monograph on non-standard packing problems, with additional conditions, also of 'transversal' nature (e.g. balancing), relies on a modelling-based GO approach. This point of view, espoused for the tetris-like item orthogonal packing, can be extended, at least at an approximate level, to more complex frameworks. This holds, in particular, for objects like polyhedrons, with the possibility of continuous rotations. According to the approach put forward here, the tetris-like formulation still plays an important role, providing a 'naïve' starting global solution.

The literature on the packing of complex (non-rectangular) objects is extensive (e.g. Bennell and Oliveira 2008; Betke and Henk 2000; Cagan et al. 2002; Chernov et al. 2010; Egeblad et al. 2009a, b; Gan et al. 2004; Kallrath 2009; Torquato and Jiao 2009), also including quite sophisticated formulations, but mostly addressed to local optimization. A methodology of particular interest in this sense (Scheithauer et al. 2005; Stoyan and Chugay 2009; Stoyan et al. 1996, 2012) could well serve the scope of improving the approximate (quasi-global) solutions obtained with the approach discussed in this section. Stoyan's method introduces the concept of Φ-functions (e.g. Chernov et al. 2012; Stoyan et al. 2002, 2004). These are briefly outlined here below (limiting the discussion, for simplicity, to the two-dimensional case without rotations).

Given two general items $A_i(\boldsymbol{o}_i)$ and $A_j(\boldsymbol{o}_j)$, where $\boldsymbol{o}_i = (o_{1i}, o_{2i})$ and $\boldsymbol{o}_j = (o_{1j}, o_{2j})$ represent their local reference frame position, respectively, any everywhere continuous function $\Phi_{ij} : \boldsymbol{R}^4 \to \boldsymbol{R}$ is called a Φ-function of $A_i(\boldsymbol{o}_i)$ and $A_j(\boldsymbol{o}_j)$ if it possesses the following properties:

$\Phi_{ij} > 0$ if $A_i(\boldsymbol{o}_i) \cap A_j(\boldsymbol{o}_j) = \{\emptyset\}$
$\Phi_{ij} = 0$ if int $A_i(\boldsymbol{o}_i) \cap$ int $A_j(\boldsymbol{o}_j) = \{\emptyset\}$ and $\partial A_i(\boldsymbol{o}_i) \cap \partial A_j(\boldsymbol{o}_j) \neq \{\emptyset\}$
$\Phi_{ij} < 0$ if int $A_i(\boldsymbol{o}_i) \cap$ int $A_j(\boldsymbol{o}_j) \neq \{\emptyset\}$

In such a way, $\Phi_{ij} \geq 0$ guarantees that items $A_i(\boldsymbol{o}_i)$ and $A_j(\boldsymbol{o}_j)$ do not intersect (apart from their borders).

This section discusses the (two-dimensional) placement of simple polygons, i.e. polygons with no intersection between two nonconsecutive edges, inside a convex polygon. The approach introduced is closer, with respect to Stoyan's one, to alternative GO-based methodologies (e.g. Fischetti and Luzzi 2009; Sykora et al. 2011). An MINLP model is formulated. It is intended to be processed recursively, following a successive approximation philosophy. Once an acceptable (approximate) solution has been obtained, it can be exploited, as a starting point to solve the corresponding exact Φ-function-based MINLP model. The reader is referred to the previous work (Fasano 2013), for more details.

6.2.1 Approximate MINLP Model

Some necessary conditions, formulated in terms of MINLP constraints, are considered hereinafter. They are aimed at looking into approximate solutions to the two-dimensional problem of placing simple polygons (in the following just called polygons) from a given set I_P, into a convex polygon D (domain). The overall surface of the loaded items is maximized. For each polygon, any possible orientation is admitted. A recursive process is performed to improve, by successive approximation, the current solution, until a satisfactory one is reached. The positioning rules for each picked item are simply:

- *Each polygon has to be contained within D (domain conditions).*
- *Polygons cannot overlap (non-intersection conditions).*

To formulate the corresponding mathematical model, we shall consider a given (main) reference frame with origin O and axes w_β, $\beta \in \{1, 2\}$ (the set of axes is still denoted by B, cf. Sect. 2.1). The domain D is delimited by the set of vertices V_P, whose coordinates, with respect to the main reference frame, are represented by $V_{P\beta\gamma}$, $\gamma \in V_P$. They are assumed as nonnegative (without loss of generality). We shall then consider any polygon i (denoted in the following by P_i), from the given set I_P, and associate to it a local reference frame with origin $\boldsymbol{o}_{Pi}$, of coordinates $o_{P\beta i}$ (with respect to the main reference frame). The set of all vertices associated to polygon i is denoted by E_{Pi}. The coordinates of each vertex $\eta \in E_{Pi}$ are indicated, with respect to the local reference frame, by $V_{P\beta\eta i}$. The vector equations below hold

$$\forall i \in I_P, \forall \eta \in E_{Pi} \quad \boldsymbol{w}_{P\eta i} = \chi_{Pi} \boldsymbol{o}_{Pi} + \chi_{Pi} \left\| q_{\beta\beta'} \right\|_i V_{P\eta i}. \tag{6.16}$$

Here for each vertex $\eta \in E_{Pi}$, $\boldsymbol{w}_{P\eta i} = (w_{P1\eta i}, w_{P2\eta i})^T$ is the vector of its coordinates with respect to the main reference frame; $\boldsymbol{o}_{Pi} = (o_{P1i}, o_{P2i})^T$; $V_{P\eta i} = (V_{P1\eta i}, V_{P2\eta i})^T$; $\left\| q_{\beta\beta'} \right\|_i$ is the (orthogonal) rotation matrix of the local reference frame, with respect to the main one, and $\chi_{Pi} \in \{0, 1\}$, as in the previous cases, has the meaning: $\chi_{Pi} = 1$ if polygon i is picked; $\chi_{Pi} = 0$ otherwise.

The *domain* conditions below are stated to guarantee that each picked polygon i lies within the given polygon D:

$$\forall \beta \in B, \forall i \in I_P, \forall \eta \in E_{Pi}$$
$$w_{P\beta\eta i} = \sum_{\gamma \in V_P} V_{P\beta\gamma} \lambda_{P\gamma\eta i}, \tag{6.17}$$

$$\forall i \in I_P, \forall \eta \in E_{Pi} \quad \sum_{\gamma \in V_P} \lambda_{P\gamma\eta i} = \chi_{Pi}, \tag{6.18}$$

where the variables λ are nonnegative and have the same meaning as in Sect. 2.1.

While in the case of tetris-like items, the *non-intersection* conditions are quite easy to state, dealing with polygons they become much more complex. Three easy-to-prove necessary conditions are posed hereinafter. They establish a basis for the recursive process proposed (that acts by successive approximation). The following propositions are then stated.

Proposition 6.1 *Given a set of internal circles C_{Pi} and C_{Pj}, for any pair of polygons i and j, respectively, no circle of C_{Pi} can intersect a circle of C_{Pj}.*

Proposition 6.2 *For any pair of polygons i and j, no vertex of P_i can belong to any circle of C_{Pj} and vice versa.*

Proposition 6.3 *For each pair of polygons i and j, any set of points of P_i must belong to the external region of P_j and vice versa: this holds in particular for all vertices of the polygons.*

Remark 6.4 In the above propositions, tangency conditions are admitted. In particular, it is understood that the external regions enclose the respective boundaries.

The third necessary *non-intersection* conditions posed above (Prop. 6.3) can be advantageously restricted to bounded external *slices*. To this purpose, the concept of *augmented polygon* is introduced by the following definition.

Definition 6.1 (Augmented polygon) For each P_i, consider the polygon, denoted by $\overline{P}_i$, such that

Fig. 6.3 Example of
augmented polygon

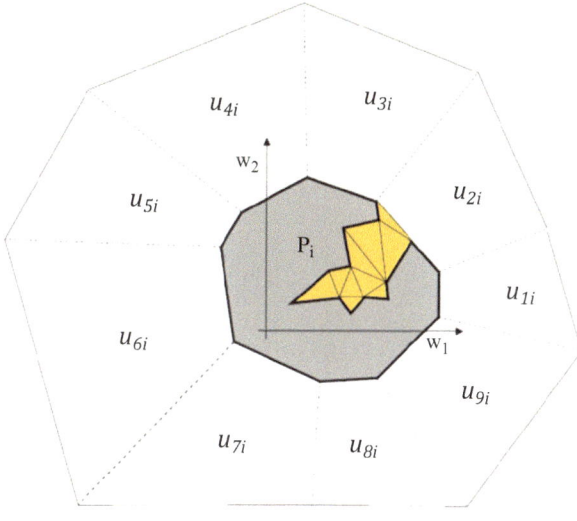

$$P_i \subset \overline{P}_i,$$

$$\overline{P}_i - P_i = \bigcup_{\nu \in \overline{S}_i} U_{\nu i},$$

where $U_{\nu i}$ (slices) are convex polygons (not necessarily disjoint), associated to P_i, and $\overline{S}_i$ is their set. Each $\overline{P}_i$ is called augmented polygon associated to P_i ($\overline{P}_i - P_i$ could always be partitioned into a set of triangles; see Fig. 6.3 and, for instance, de Berg et al. (2000) for polygon triangulation).

Figure 6.3 shows, as a matter of example, how a simple polygon can be augmented by convex slices. It is moreover immediately seen that whenever an internal 'cleft' (consisting of a non-convex simple polygon) is present, it can be partitioned into a set of triangles (i.e. convex *slices*). A specific case of the augmented polygon concept, adopted in the following, is provided by the definition below.

Definition 6.2 (Domain-covering augmented polygon) For each polygon i, any associated augmented polygon that covers the whole domain D, for any position and orientation of i within D, is called domain-covering augmented polygon, associated to polygon i. It is denoted by $\overline{\overline{P}}_i$.

The third necessary *non-intersection* conditions (Prop. 6.3), when restricted to bounded external regions, can therefore simply be expressed as follows:

For each pair of polygons i and j, with any associated $\overline{\overline{P}}_i$ and $\overline{\overline{P}}_j$, each point of P_i must belong to $\overline{\overline{P}}_j$ and, vice versa, each point of P_j must belong to $\overline{\overline{P}}_i$.

It is then immediately seen that Proposition 6.1 is expressed by the following constraints:

$$\forall i,j \in I_P / i < j, \forall h \in C_{Pi}, \forall k \in C_{Pj}$$

$$\sum_{\beta \in B} \left(o_{P\beta hi} - o_{P\beta kj} \right)^2 \geq \chi_{Pij} \left(R_{hi} + R_{kj} \right)^2. \tag{6.19}$$

Here C_{Pi} and C_{Pj} denote the (arbitrary) sets of internal circles associated to polygons i and R_{hi} and R_{kj} the radius of the relative circles; $o_{P\beta hi}$ and $o_{P\beta kj}$ are their centre coordinates, with respect to the main reference frame; and the (implicitly binary) variables $\chi_{Pij} \in [0, 1]$ are subject to the same constraints expressed by (2.12a), (2.12b) and (2.13). The following vector equations hold:

$$\forall i \in I_P, \forall h \in C_{Pi}$$

$$\boldsymbol{o}_{Phi} = \chi_{Pi}\boldsymbol{o}_{Pi} + \chi_{Pi} \left\| \boldsymbol{q}_{\beta\beta'} \right\|_i \boldsymbol{O}_{Phi}. \tag{6.20}$$

They represent (with obvious meaning of the symbols), for the centre of circle h, the coordinate transformation between the local reference frame (associated to polygon i) and the main one. Proposition 6.2 is very similar and it is not reported.

Given a *domain-covering augmented polygon* $\overline{\overline{P}}_i$, associated to polygon i, the corresponding set of *slices* are denoted by $\overline{\overline{S}}_i$. The set of vertices delimiting each *slice* ν (of $\overline{\overline{P}}_i$) is instead represented by $\overline{\overline{E}}_{\nu i}, \nu \in \overline{\overline{S}}_i$. The following constraints express, for the polygon vertices, the third necessary *non-intersection* conditions (Prop. 6.3):

$$\forall \beta \in B, \forall i,j \in I_P, \forall \eta \in E_{Pi}$$

$$\chi_{Pij} w_{P\beta\eta i} = \sum_{\substack{\gamma \in \overline{\overline{E}}_{\nu j}, \\ \nu \in \overline{\overline{S}}_j}} \lambda_{P\eta i \gamma \nu j} \, w_{P\beta\gamma\nu j}, \tag{6.21}$$

$$\forall i,j \in I_P, \forall \eta \in E_{Pi}, \forall \nu \in \overline{\overline{S}}_j$$

$$\sum_{\gamma \in \overline{\overline{E}}_{\nu j}} \lambda_{P\eta i \gamma \nu j} = \chi_{P\eta i \nu j}, \tag{6.22}$$

$$\forall i,j \in I_P, \forall \eta \in E_{Pi} \quad \sum_{\nu \in \overline{\overline{S}}_j} \chi_{P\eta i \nu j} = \chi_{Pij}, \tag{6.23}$$

where, as before, $w_{P\beta\eta i}$ are the coordinates of polygon i vertices with respect to the main reference frame. Similarly, $w_{P\beta\gamma\nu j}$ are the vertex coordinates of *slices* ν associated to polygon j; $\lambda_{P\eta i \gamma \nu j}$ are nonnegative variables and $\chi_{P\eta i \nu j} \in \{0, 1\}$. Constraints (6.21), (6.22) and (6.23) ensure thus that if both polygons i and j are loaded, then each vertex of polygon i will belong to one *slice* ν of (the *augmented polygon* associated to) j and vice versa.

Remark 6.5 As is easily gathered, the presence of the $\chi_{P\eta i \nu j}$ binary variables increases the model complexity dramatically. Constraints (6.22) and (6.23) could thus be profitably substituted by the following $\forall i,j \in I_P, \forall \eta \in E_{Pi}, \forall \nu \in \overline{\overline{S}}_j \quad \sum_{\gamma \in \overline{\overline{E}}_{\nu j}} \lambda_{P\eta i \gamma \nu j} = \chi_{Pij}.$

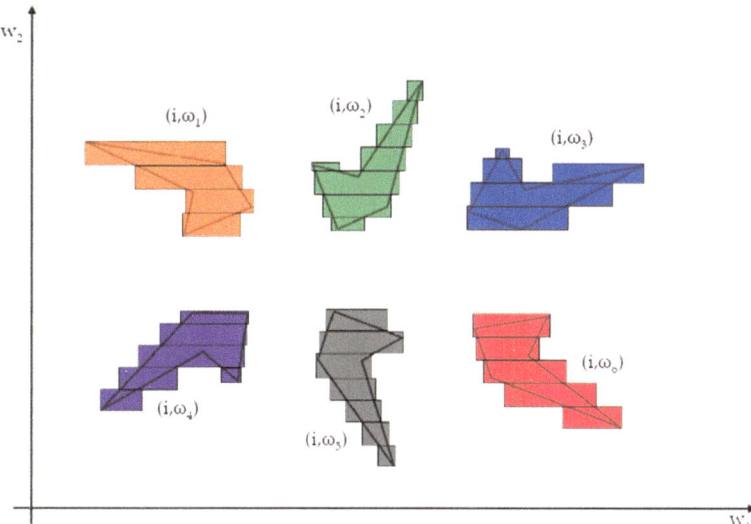

Fig. 6.4 Covering tetris-like items (corresponding to possible rotations $\omega_1 - \omega_6$, of the same polygon i)

The logical restriction expressed by constraints (6.23) may indeed be treated algorithmically, by introducing appropriate special ordered sets, similar to those suggested by Escudero (1988). More specifically, in such a case, only the variables $\lambda_{P\eta i\gamma\nu j}$ corresponding to a single slice $\nu \in \overline{\overline{S}}_j$ would be allowed to be positive, while all remaining are forced to zero.

6.2.2   Applications

The difficulty in solving the MINLP model of Sect. 6.2.1 is, *per se*, extremely high, even when small-scale instances are involved. Consequently, it is not expected to accomplish the task directly, and an incremental procedure is strongly recommended. As a rough approximation, for instance, for each polygon i, just one of its biggest internal circles could be considered. The number of internal circles could be sequentially increased, for all pairs of polygons currently intersecting, until a satisfactory (approximate) solution is attained. And similar considerations hold for all the necessary conditions considered in Sect. 6.2.1.

As, in any case, the MINLP solution process efficiency is strongly affected by the initial guess available, the tetris-like model of Sect. 2.1 can be utilized for this purpose. This can be done by temporarily replacing the given polygons with covering tetris-like items and considering, for each polygon $i_P \in I_P$, just a set Ω_{Pi} of possible (arbitrary) discretized rotations; see Fig. 6.4.

For each polygon $i_P \in I_P$ and each selected rotation, $\omega_{Pi} \in \Omega_{Pi}$, let us define a single tetris-like item, covering the polygon for that rotation and such that its sides are orthogonal/parallel to the main reference frame axes. Denoting by I_T the set of all tetris-like items built in such a way, the one-to-one correspondence $(i_P, \omega_{Pi}) \in I_P \times \Omega_{Pi} \leftrightarrow i_T \in I_T$ is defined. This way, the subset of tetris-like items I_{Ti_P} is associated to each polygon $i_P \in I_P$. The problem in question is hence that of placing the covering tetris-like items of set I_T, one (and only one) for each subset I_{Ti_P}, into D, without any possibility of rotation. This leads to a special case of the MIP model of Sect. 2.1: constraints (2.1) are dropped, whilst (2.2), (2.3), (2.4), (2.5a), (2.5b) and (2.6) are kept, setting all variables θ to one and eliminating all indexes ω, as well as the related sums. Let us denote (with a little of abuse of notation) by $\chi_{i_P} \in \{0, 1\}$ and $\chi_{i_T} \in \{0, 1\}$ the decisional variables controlling the selection of polygon $i_P \in I_P$ and its associated (pre-oriented) tetris-like item $i_T \in I_{Ti_P}$. The following conditions have to be added:

$$\forall i_P \in I_P \quad \sum_{i_T \in I_{Ti_P}} \chi_{i_T} = \chi_{i_P}. \tag{6.24}$$

This guarantees that if polygon i_P is picked, it is represented by one and only one tetris-like item i_T, corresponding to a specific orientation (from the set of the discretized ones associated to i_P).

Remark 6.6 An alternative tetris-like item approximation could be considered, keeping all the constraints (2.1), (2.2), (2.3), (2.4), (2.5a), (2.5b) and (2.6) (formally) as they are. Each projection $L_{\omega\beta hi}$ would be associated to a specific (discretized) orientation of the related polygon. These terms, however, would no longer correspond to rigid rotations of the relative tetris-like items (covering tetris-like items, indeed, change their pattern, depending on the orientation). Figure 6.4 clearly illustrates this aspect.

With the approximation approach suggested, any tetris-like item covers the corresponding polygon for a specific rotation of it. Any feasible solution of the tetris-like item problem is also a feasible solution of the polygon problem and thus represents (for the surface maximization) a lower bound. Once a good initial solution is attained, the actual items (polygons) can be introduced and the MINLP process activated.

A heuristic dedicated to the polygon packing problem is currently at a prototyping stage (see Fasano 2013; LGO Solver Suite for Global–local Nonlinear Optimization is utilized as a nonlinear optimizer, see Pintér 1997, 2002, 2005, 2007; Pintér Consulting Services 2013). It performs the tetris-like approximation as an initialization step (*virtual* items are purposely introduced, in order to concentrate unexploited areas in a limited number of uncovered zones). Then, different general techniques, such as item fixing/exchange and 'hole' filling, are adopted (exploiting the features of Prop. 6.1, 6.2 and 6.34). An experimental investigation is currently under study (Fasano 2013).

Concerning the computational aspects related to the first phase, based on a two-dimensional tetris-like item MIP model, insights can be derived from Chap. 5.

Fig. 6.5 Domain with internal polygonal 'holes'

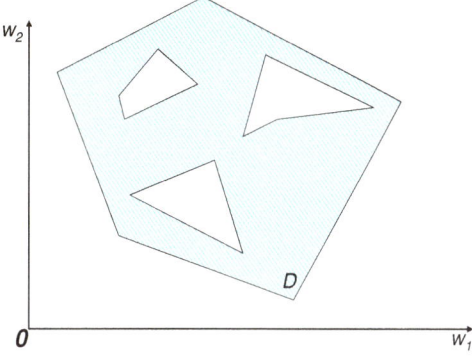

Fig. 6.6 Simple polygons with 'holes'

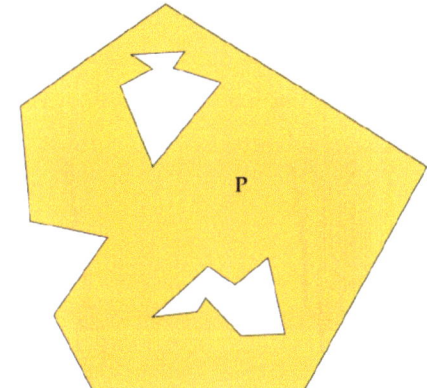

The necessary conditions expressed by Propositions (6.1) and (6.2) rely, instead, on the MINLP formulation of *circle packing*, and significant topical literature is available (e.g. Castillo et al. 2008; Hifi and M'Hallah 2009; Specht 2012).

It is argued that the additional conditions discussed in Sect. 2.1 could also be (at least in principle) extended to the case of polygon packing. This holds, for instance, when balancing restrictions are stated. Domains with 'holes' or forbidden zones can in general be modelled by introducing zero-mass items (see Fig. 6.5). Moreover, when some items contain 'holes', these become part of their external regions (see Fig. 6.6).

Chapter 7
Directions for Future Research

Although this volume summarizes results derived from more than 2 decades of dedicated research, a number of study directions still remain open. This is mainly due, on one side, to the heuristic nature of the overall point of view adopted that, as such, relies on a huge amount of experimental activity. The modeling philosophy followed is, moreover, subject to a wide range of possible extensions. It is then worth looking into, at least in perspective, further applications and topical formulations. Section 7.1 outlines the experimental aspects, whilst Sect. 7.2 focuses on the modeling ones.

7.1 Experimental Context

An extensive testing procedure has been carried out, as reported in Sect. 5.5, involving standard instances from literature, i.e. concerning the orthogonal packing of parallelepipeds, into a parallelepiped, with no additional conditions.

The non-standard packing problems considered in this work can hardly be classified by following a systematic scheme and, therefore, also the relevant experimental analysis is very demanding. The efficiency of the heuristic approaches proposed in Chap. 4 is affected by a variety of factors. The difficulties of the instances to solve, indeed, not only depend on the number of items/*components* involved. They are also strongly related to the characteristics of the objects, the domain geometry and the presence of additional conditions.

The *separation* planes significantly reduce the volume exploitation. It is, moreover, understood that the tighter the centre of mass domain is, the tougher the problem becomes. Roughly speaking, it could be said that the (*static*) balancing conditions, with some percentage of admissible off-centring (with respect to the container dimensions), can decrease the volume exploitation by 15–20 % and increase the computational effort by up to 25–30 %. These estimates are, however, very imprecise and indicate just a general rule of thumb.

G. Fasano, *Solving Non-standard Packing Problems by Global Optimization and Heuristics*, SpringerBriefs in Optimization, DOI 10.1007/978-3-319-05005-8_7, © Giorgio Fasano 2014

Extensive statistics based on a tentative (rough) classification of problems involving tetris-like items and (differently combined) additional conditions represents, without any doubt, quite a demanding research objective for the future. The experience acquired to date and referred to in this book, indeed, has been confined to the packing of tetris-like items, inside curved domains, with *separation* planes, subject to (usually quite tight) balancing conditions. The reason is due to the fact that the research surveyed here has been mainly motivated by the space engineering context.

An in-depth testing of further additional conditions would therefore represent a first important step to widen the relevant knowledge regarding the computational aspects. Real-world instances, arising in different engineering and logistics fields, would certainly enrich the overall statistical information, even if this interdisciplinary task represents a remarkable difficulty.

A further experimental investigation related to the *feasibility* subproblem is currently ongoing. A significant effort to consolidate and, hopefully, even improve the present outcomes is expected. This holds, in particular, both for the second linear reformulation of Sect. 3.1.2 (including its variation, also reported there) and the nonlinear one of Sect. 3.1.4 that seem promising.

As previously pointed out, the non-restrictive reformulation of the general Mixed Integer Programming (MIP) model of Sect. 3.1.3 is, per se, worthy of a dedicated in-depth experimental analysis. Moreover, its possible utilization, to support the *initialization* step of Sect. 4.3.1 (see Remark 4.9), definitely deserves further attention. Similar considerations concern the adoption of this reformulation as an alternative to the *item-exchange* and *hole-filling* modules (see Sect. 4.3.3).

The implications and *valid* inequalities overviewed in Sect. 3.2 (and possible further formulations) would deserve dedicated experimentation. Their exhaustive generation can hardly be carried out when a *branch-and-bound* approach is adopted, even for small-scale instances. An ad hoc methodology to select limited subsets of such auxiliary constraints could thus be investigated. This way, the model could be *tightened*, without making the instance intractable from the dimensional point of view. It is, moreover, understood that an experience-based generation process, specifically oriented to a *branch-and-cut* approach, would also be of great interest. It is however expected to be quite demanding, both in terms of development and testing.

In any case, independently from the general algorithms adopted to solve the various models (e.g. *branch-and-bound*), a crucial aspect concerns the search strategy (for instance, in terms of branching sequence, variable processing priority, use of preprocessing/probing/rounding/fixing techniques and so on, e.g., Chen et al. 2010, Linderoth and Savelsbergh 1999, Marchand and Wolsey 1998). This holds in general, both for the MIP and Mixed Integer Non-Linear Programming (MINLP) models considered, so that the way the relative solvers are driven can be very influential on the computational efficiency. A dedicated experimental activity is currently being carried out.

Whilst improvements in terms of single model solution, as mentioned above, are expected, an experimental 'tuning' of the heuristic overall processes represents a further very important objective. Indeed, the heuristics presented in this monograph are based on several arbitrary choices. In the process of Sect. 4.3.2, for instance, the dichotomy-based partitioning of the given set of items can be performed following different strategies. A similar situation of arbitrary choice arises when executing the *item-exchange* step of Sect. 4.3.3.

The stopping rules established (a priori) for each MIP/MINLP solution process are usually very influential, as well as those concerning the item prioritization. A wide-range statistical analysis could well suggest how to achieve a dynamic setting of the relevant parameters that best match, each time, the specific instance to solve. A dedicated research activity should therefore be addressed to these aspects.

The two heuristic approaches proposed in Chap. 4, moreover, have their pros and cons, depending on the framework involved and, often, it is hardly understood a priori which one represents, case by case, the best choice. This shows that a hybrid system, based on the joint use of both, could be desirable in most practical situations. Also the combined utilization with other packing algorithms (in addition to the already mentioned non-restrictive reformulation of the general MIP model), especially in the *initialization* phase, could be of interest. All this paves the way to an experimental analysis in this perspective.

Even hybridizations at model level could be taken into account. A merging of the MIP model utilized by the heuristic of Sect. 4.3.2 and that of Sect. 3.1.2 could be, for instance, implemented, introducing 'proper' weights in the resulting (combined) *objective* function. The contribution of Sect. 3.1.2 model could act, this way, in support to the heuristic process as an 'accelerator factor'. A dedicated experimentation would thus represent an interesting research topic.

Concerning the issue of exploiting the empty spaces, by adding *virtual* items, extensive computational outcomes relative to the logarithmic formulation discussed in Sect. 6.1.2 are desirable. Also the use of both the *surrogate objective* functions, proposed there, to provide an initial approximate solution to the MINLP model of Sect. 6.1.1 would be of interest.

As previously pointed out, the heuristic, briefly mentioned in Sect. 6.2.2, is currently at a preliminary stage. A significant experimentation is foreseen to properly tune all the relevant steps, including the one based on the tetris-like item first approximation. The use of the Φ-functions (cf. Sect. 6.2) to refine (by local optimization) the approximate solutions obtained with the approach proposed is matter of promising further investigations.

In addition to the experimental research directions suggested so far, it is still worth considering a further application of the general MIP model second linear reformulation, in the *LP-relaxed* version put forward in Sect. 4.3.1.2. This could indeed serve the scope of providing 'good' initial guesses in support of circle/ sphere packing, when a global optimization approach is needed. To this purpose, expressions (4.2) have to be properly restated. For each given circle (/sphere), the lower bound appearing there is substituted with the side of the relative inscribed square (/cube). Analogously, the side of the circumscribed square (/cube) replaces the corresponding upper bound. This way, an approximate solution, consisting of the placement of rectangles (/parallelepipeds), inside the given domain is looked for (each of them encloses the corresponding inscribed square (/cube) and is included in the circumscribed one).

7.2   Modeling Enhancements and New Applications

7.2.1   Extension of the Packing Models

In Sect. 2.3 a number of possible additional conditions have been considered. Nonetheless, this is an area susceptible to a wide range of possible extensions, taking into account different applicative sectors, each being characterized by specific requirements. The approach put forward, indeed, is quite flexible and suitable for including, whenever necessary, the appropriate additional conditions, without modifying the overall structure of the model.

Further research is foreseen to provide new nonlinear reformulations alternative to the ones presented in Sect. 3.1.4, as well as to identify further implications and *valid* inequalities.

Enhancements concerning the heuristic of Sect. 4.3.3 could represent the objective of future developments. This holds, in particular, for the *hole-filling* phase. A new procedure, based on the generation of *virtual* items, is intended to be activated prior to it, in order to 'enlarge' the 'holes' present (preparing, in such a way, the room for items not yet loaded). To this purpose, an appropriate combination of the *packing* model of Sect. 4.3.3 and the *virtual* item one of Sect. 6.1 is currently under study.

Further development concerning this heuristic could, moreover, be stressed. Indeed, when the *packing* model is solved, after the *hole-filling* step execution, items are tentatively inserted, considering all possible interactions with those already loaded. This makes the solution process quite time consuming because of the *non-intersection* constraints. In order to reduce their number, the concept of *virtual cage* is to be introduced (see Fig. 7.1). It consists of a 'virtual'

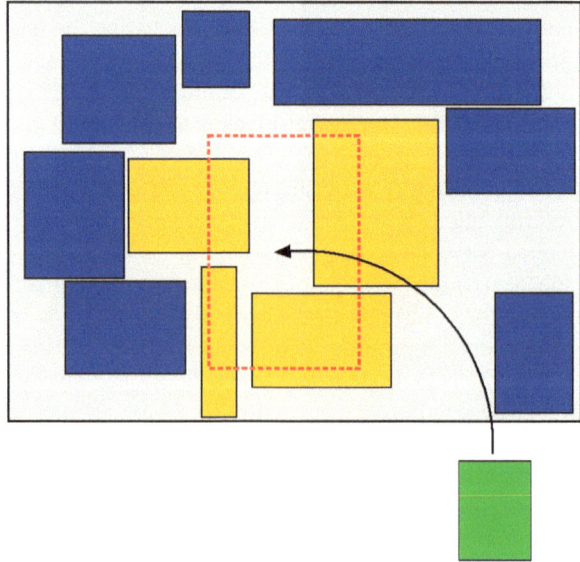

Fig. 7.1 *Virtual cage* basic concept

Fig. 7.2 Case of
non-simple polygons

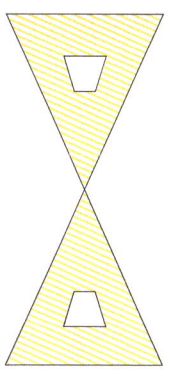

parallelepiped, associated to each item that is expected to be added, and centred
with respect to the corresponding grid node, as selected by the *hole-filling* step. The
non-intersection conditions (in the *packing* model), involving the prospective
additional item, are then limited only to the ones that (in the *hole-filling* solution)
intersect the corresponding *virtual cage*. The remaining items are forced to main-
tain the *non-intersection* state with respect to the *virtual cage* itself.

The *global optimization* approach discussed in Sect. 6.2 to tackle the polygon
packing issue is currently under study and several enhancements are foreseen. The
present formulation is limited to the case of simple polygons. It is easily seen,
however, that straightforward extensions can be carried out to include a wider class
of objects (Fig. 7.2 shows a case of a non-simple polygon that could be included
with the appropriate modeling).

It is, furthermore, obvious that the extension to the three-dimensional case,
i.e. involving polyhedrons, represents the natural evolution of the present MINLP
model.

In addition to what it is recalled above, it should be noticed that the tetris-like
item approximation, adopted to generate a starting solution for the MINLP process
(Sect. 6.2.2), relies on arbitrary choices. Firstly, it is decided what discretized
orientations have to be considered for each polygon (this determines the sets
Ω_{Pi}). Secondly, a tetris-like item is arbitrarily generated for each polygon and
each orientation of it. This point itself could be the subject of a dedicated optimization problem, stated as follows:

*Given a polygon (with prefixed orientation in an orthogonal reference frame),
let us define a covering tetris-like item (oriented orthogonally with respect to
the same frame) of minimum surface, consisting of an established number of
components.*

These kinds of approximating tetris-like items could well be automatically
generated by a dedicated optimization process, based on ad hoc models/algorithms.
This aspect also represents quite a challenging objective of future developments.

7.2.2   Application to Scheduling Problems
with Nonconstant Resource Availability
and Nonconstant Operational Cycles

The application considered next is motivated by a real-world scenario. It is relevant to the International Space Station (ISS, cf. http://www.nasa.gov) context, where the resource availability is scarce, whilst the demand, to perform the necessary on-board activity, quite high. The related optimization framework is of remarkable interest and even more challenging situations are expected for the space exploration programs of the near future. In this section we shall address, in particular, a demanding problem of scheduling. It consists of the activation of a number of devices, in a context of nonconstant resource availability and nonconstant operational cycles. Here the issue is dealt with in a streamlined version, with the confidence, however, that the topic can be of interest for a wide range of applications, albeit in very different contexts.

Just to focus on an exemplificative instance, let us consider the case of electrical power as the only resource involved. We shall assume that a certain number of devices have to perform series of cycles, in a given period of time. The relevant optimization problem consists of scheduling the activation of each device-cycle, so that a certain profit criterion is maximized (e.g. utilizing as much of the energy available as possible). Obviously several operational conditions, affecting the activation of the various cycles, could be present.

If the electrical power available were constant, as well as the consumption of each device-cycle, then the relevant optimization problem would be equivalent to that of packing rectangles into a rectangle (on possible interrelations between packing and scheduling problems consult, for instance, Alvim and Ribeiro 2004; Chiong and Dhakal 2009; Liu and Baskiyar 2008; Zhang 2004). Indeed, in such a case, each cycle would be characterized by its duration and its power consumption. The area of the corresponding rectangle would simply represent the energy consumption associated to each cycle. Similarly, the rectangle having the given time period as a base and the (constant) power availability as a height, in a power-time reference frame, would represent the total amount of energy.

A much more complex situation occurs, however, when the power availability is not constant and/or the same holds for the consumption associated to each cycle. Figure 7.3 shows an example of the operational scenario under consideration here.

The situation illustrated is quite similar to that of packing two-dimensional tetris-like items (corresponding to cycles with nonconstant power consumption) inside a stepwise two-dimensional domain (corresponding to the nonconstant power availability over the whole time period). Some important peculiarities have, however, to be pointed out. Firstly, it is understood that in this context, the tetris-like items, since they represent operational cycles, cannot be rotated.

Let us suppose, then, that, for a certain device, the relative cycles must be executed at a fixed rate. Assume, moreover, that the number of cycles itself is fixed, so that either all of them (i.e. no more and no less) are performed, in the given

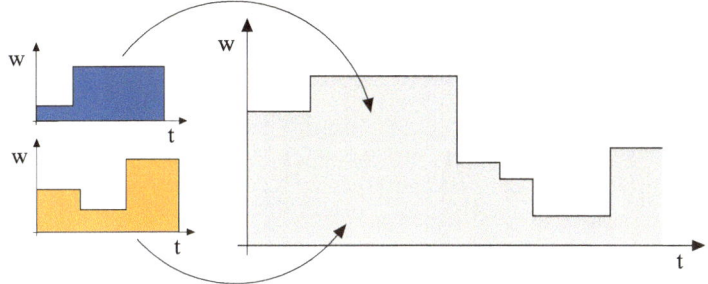

Fig. 7.3 Nonconstant resource availability and nonconstant operational cycles

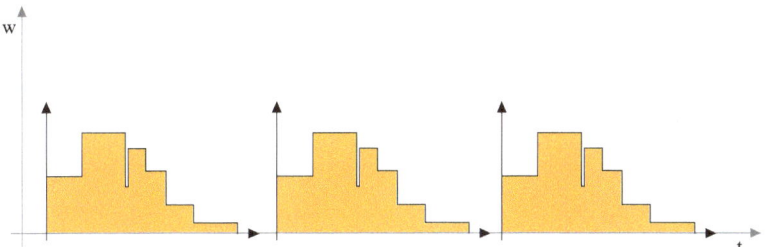

Fig. 7.4 *Disconnected* tetris-like item

Fig. 7.5 Illustrative example

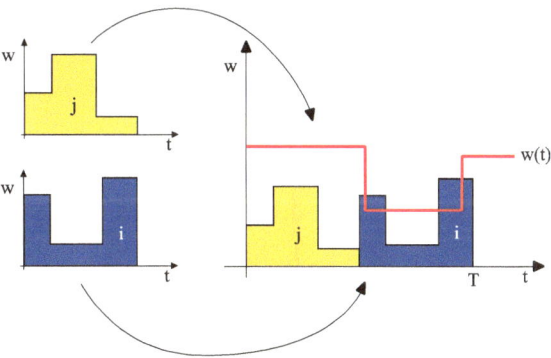

time period, or the device is not activated. This suggests that we should interpret the whole sequence of cycles, relative to such a device, as a single tetris-like item, as depicted in Fig. 7.4. It is denoted as a *disconnected* tetris-like item.

We shall consider, now, the example illustrated in Fig. 7.5. Two devices i, j are expected to execute just a single cycle each within the time period $[0, T]$. The given electrical power (step) function is denoted by $w(t)$, where $t \in [0, T]$ represents the time variable.

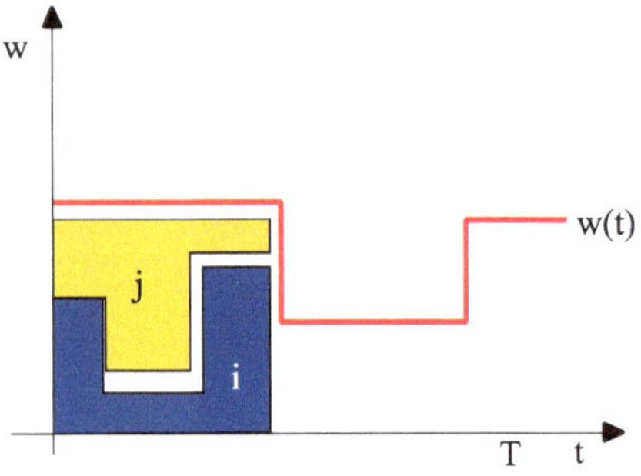

Fig. 7.6 Illustrative
example optimal solution

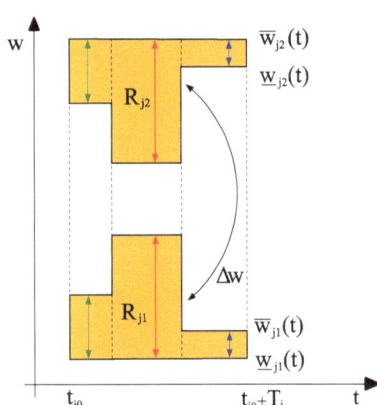

Fig. 7.7 Example
of different representations
of the same cycle

As seen in Fig. 7.5, if the cycles are represented by the tetris-like items depicted
on the left, at most one of them can be activated (indeed, only one of the associated
tetris-like items can stay inside the domain). The optimal result, however, is that
illustrated in Fig. 7.6, where both devices are working.

In this case, in fact, the cycle associated to device j has been provided with a
different representation. To see that it is equivalent to the previous one, let us
consider the general diagram (t,w) of Fig. 7.7, where the two representations R_{j1} and
R_{j2} of the same (single) cycle, associated to device j, are shown.

Each of them is defined by the two functions $\underline{w}_{j1}(t)$, $\overline{w}_{j1}(t)$ and $\underline{w}_{j2}(t)$, $\overline{w}_{j2}(t)$,
respectively ($\underline{w}_{j1}(t)$ and $\underline{w}_{j2}(t)$ are the lower functions whilst $\overline{w}_{j1}(t)$ and $\overline{w}_{j2}(t)$ the
upper ones). All these functions are zero in the whole interval $[0, T]$ except for
$t \in [t_{j0}, t_{j0} + T_j]$, where t_{j0} is the instant when the cycle is activated and T_j is its
duration (see Fig. 7.7):

$$\forall t \in [0, T]$$
$$\overline{w}_{j1}(t) - \underline{w}_{j1}(t) = \overline{w}_{j2}(t) - \underline{w}_{j2}(t).$$

This clearly means that the power consumption associated with device j at any
instant $t \in [0, T]$ is $\Delta w_j(t) = \overline{w}_{j1}(t) - \underline{w}_{j1}(t)$ or $\Delta w_j(t) = \overline{w}_{j2}(t) - \underline{w}_{j2}(t)$ and the

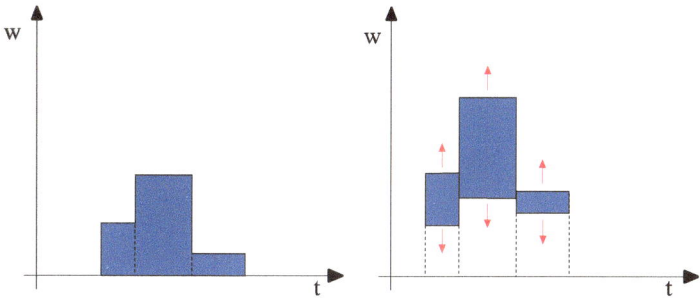

Fig. 7.8 *Non-rigid* tetris-like item basic concept

two representations R_{j1} and R_{j2} are equivalent (the heights of the functions $\underline{w}_{j1}(t)$ and $\underline{w}_{j2}(t)$ are, with respect to the time axis, obviously arbitrary. Instead, $\Delta w_j(t)$, during the whole working time period $t \in [t_{j0}, t_{j0} + T_j]$, must exactly represent the consumption associated to device j). In the solution illustrated in Fig. 7.6, the *feasibility* is guaranteed by the compliance with the following condition:

$$\forall t \in [0, T] \quad \Delta w_i(t) + \Delta w_j(t) \leq w(t).$$

All this suggests that the concept of *non-rigid* tetris-like item should be introduced to contemplate any possible cycle representation (in the sense clarified above). We assume therefore that each *non-rigid* tetris-like item consists of a set of *components* that are fixed with respect to the time axis. They can however be arbitrarily translated along the vertical one (corresponding to the resource that, in our specific case, is the electrical power). It is obviously assumed that the *components* belonging to the same *non-rigid* tetris-like item cannot overlap (apart from their boundaries). As already pointed out, moreover, items (being representations of cycles) are not allowed to rotate. The concept is illustrated in Fig. 7.8.

The model formulation of Sect. 2.1 can easily be adapted to the scheduling scenario in question and the concept of *non-rigid* tetris-like item just introduced. The main changes are outlined next, supposing, for the sake of simplicity, that each device can perform just a single cycle (a generalization of this reduced case is straightforward, also when *disconnected* tetris-like items are involved).

Each device-cycle is therefore associated to a *non-rigid* tetris-like item and a main orthogonal reference frame (t, w) is introduced (limiting its variables to the first quadrant only). As no rotation is admitted, the *orthogonality* conditions (2.1) are eliminated. In the following, for each device-cycle $i \in I_D$ (set of devices), t_{i0} (assumed as non-negative) shall denote its activation time, with respect to the main reference frame; T_{hi0} the starting time, with respect to t_{i0}, of its phase h (corresponding to *component* h, with $h \in C_{Di}$, set of *components* of device-cycle i) and T_{hi} its duration. Similarly, W_{hi} represents the electrical power consumption associated to phase h and supposed constant for all its duration. The binary variables $\chi_{Di} \in \{0, 1\}$ are introduced next, with the obvious meaning that

Fig. 7.9 Domain associated to a stepwise function

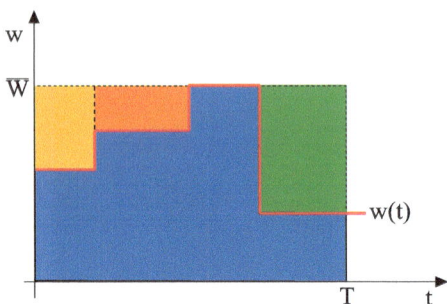

$\chi_{Di} = 1$ if the corresponding device is activated in the time period $[0, T]$ and $\chi_{Di} = 0$ otherwise.

Conditions (2.2) are then substituted by the following (see Fig. 7.8):

$$\forall i \in I_D, \forall h \in C_{Di} \quad t_{hi} = t_{i0} + \left(T_{hi0} + \frac{1}{2}T_{hi}\right)\chi_{Di}, \tag{7.1}$$

where the variables t_{hi} denote, for each *component* (i.e. phase) h of i, the coordinate t of its centre, with respect to the main reference frame. Similarly, w_{hi} (assumed as non-negative) shall indicate, for each *component* h of i, the coordinate w of its centre, with respect to the same reference frame: they are, however, no longer subject to conditions such as the ones expressed by (2.2), since each *component* is free to move vertically.

The domain (see Fig. 7.9), with origin (0,0), is determined by the time period $[0, T]$ and the stepwise function $w(t)$ defined on it. As already suggested in Sect. 2.3.3, such a domain can be delimited by introducing an enclosing rectangle of sides T and $\overline{W} = \max_{t \in [0,T]} w(t)$ that contains fixed rectangles, determining the stepwise profile of $w(t)$.

The *domain* conditions below state that if a device is activated, its associated (single) cycle must be finalized in the time period $[0, T]$. This means, in particular, that all the relative phases have to be performed, i.e.,

$$\forall i \in I_D, \forall h \in C_{Di}$$

$$\left(T_{hi0} + \frac{1}{2}T_{hi}\right)\chi_{Di} \leq t_{hi} \leq \left(T - \frac{1}{2}T_{hi}\right)\chi_{Di}. \tag{7.2}$$

The *domain* conditions corresponding to the electrical power axis are expressed by the following:

$$\forall i \in I_D, \forall h \in C_{Di}$$

$$\frac{1}{2}W_{hi}\chi_{Di} \leq w_{hi} \leq \left(\overline{W} - \frac{1}{2}W_{hi}\right)\chi_{Di}. \tag{7.3}$$

The *non-intersection* conditions (2.5a), (2.5b), (2.6), (3.15a) and (3.15b) are then substituted with the following:

$$\forall i,j \in I_D / i < j, \forall h \in C_{Di}, \forall k \in C_{Dj}$$
$$t_{hi} - t_{kj} \geq \frac{1}{2} \left(T_{hi} + T_{kj} \right) - T \left(1 - \sigma^+_{Thkij} \right), \tag{7.4a}$$

$$\forall i,j \in I_D / i < j, \forall h \in C_{Di}, \forall k \in C_{Dj}$$
$$t_{kj} - t_{hi} \geq \frac{1}{2} \left(T_{hi} + T_{kj} \right) - T \left(1 - \sigma^-_{Thkij} \right), \tag{7.4b}$$

$$\forall i,j \in I_D / i < j, \forall h \in C_{Di}, \forall k \in C_{Dj}$$
$$w_{hi} - w_{kj} \geq \frac{1}{2} \left(W_{hi} + W_{kj} \right) - \overline{W} \left(1 - \sigma^+_{Whkij} \right), \tag{7.5a}$$

$$\forall i,j \in I_D / i < j, \forall h \in C_{Di}, \forall k \in C_{Dj}$$
$$w_{kj} - w_{hi} \geq \frac{1}{2} \left(W_{hi} + W_{kj} \right) - \overline{W} \left(1 - \sigma^-_{Whkij} \right), \tag{7.5b}$$

$$\forall i,j \in I_D / i < j, \forall h \in C_{Di}, \forall k \in C_{Dj}$$
$$\sigma^+_{Thkij} + \sigma^-_{Thkij} + \sigma^+_{Whkij} + \sigma^-_{Whkij} \geq \chi_{Di} + \chi_{Dj} - 1, \tag{7.6}$$

$$\forall i,j \in I_D / i < j, \forall h \in C_{Di}, \forall k \in C_{Dj}$$
$$\sigma^+_{Thkij} + \sigma^-_{Thkij} + \sigma^+_{Whkij} + \sigma^-_{Whkij} \leq \chi_{Di}, \tag{7.7a}$$

$$\forall i,j \in I_D / i < j, \forall h \in C_{Di}, \forall k \in C_{Dj}$$
$$\sigma^+_{Thkij} + \sigma^-_{Thkij} + \sigma^+_{Whkij} + \sigma^-_{Whkij} \leq \chi_{Dj}. \tag{7.7b}$$

Further *non-intersection* conditions between the *components* and the fixed rectangles inside the domain (introduced to represent the stepwise profile of $w(t)$) should be included, but, being obvious, they are not reported here. A possible *objective* function is

$$\max_{\substack{i \in I_D, \\ h \in C_{Di}}} \sum W_{hi} T_{hi} \chi_{Di},$$

where $W_{hi} T_{hi}$ is the electrical energy consumption associated to phase h of device-cycle i.

In addition to what has been outlined up to this point, it could be shown, by simple 'fabricated' examples, that even the model based on the concept of *non-rigid* tetris-like item does not guarantee that optimal solutions are found. This is essentially due to the discretized nature of the cycle representation. A possibility to refine the relevant formulation would consist of decomposing each phase into subphases and setting the corresponding *components* free with respect to the vertical axes (obviously, the bigger the number of the subphases is, the more refined the model results). The approach proposed in this section has, as a matter of fact, just a heuristic meaning. Nonetheless, it is easily seen that any solution of it is a feasible solution also of the corresponding scheduling problem. In particular, the compliance with the necessary conditions below holds:

$$\forall t \in [0, T] \qquad \sum_{\substack{i \in I_D, h \in C_{Di}/ \\ t \geq t_{i0} + T_{hi0}, \\ t \leq t_{i0} + T_{hi0} + T_{hi}}} W_{hi} \chi_{Di} \leq w(t), \qquad (7.8)$$

$$\sum_{\substack{i \in I_D, \\ h \in C_{Di}}} W_{hi} T_{hi} \chi_{Di} \leq \int_{t \in [0,T]} w(t)\, dt. \qquad (7.9)$$

The brief discussion of this section has focused on the electrical power as the only resource involved. The extension to a more general case, where a number of different resources are considered, is quite straightforward. For each resource ρ, represented by the function $w_\rho(t)$, $\forall\ t \in [0, T]$, let us consider the plane (t, w_ρ). For each (t, w_ρ), we shall associate to each cycle, a *non-rigid* tetris-like item, corresponding to the cycle projection, with respect to the relative resource ρ (extending, as appropriate, constraints (7.3), (7.5a), (7.5b), (7.6), (7.7a) and (7.7b), to include all the relevant resources). The overall problem consists then of solving, contemporarily, for all the planes involved, the two-dimensional model described here.

As previously mentioned, the case involving also more than one single cycle for each device, in the presence of possible additional operational conditions, could quite easily be treated, on the basis of the present discussion. In addition to all this, it is expected that both *disconnected* tetris-like items and the *non-rigid* ones are susceptible to further applications, also in different fields.

It should be moreover pointed out that the heuristic approaches proposed in this monograph (Sect. 4.3) could be properly tailored to solve also the *non-rigid* tetris-like item model (provided, case by case, with the extensions needed). This certainly offers quite a motivating cue for future research and development.

Chapter 3
Model Reformulations and Tightening

G. Fasano, *Solving Non-standard Packing Problems by Global Optimization and Heuristics*, SpringerBriefs in Optimization, DOI 10.1007/978-3-319-05005-8, pp. 27–38, © Giorgio Fasano 2014

DOI 10.1007/978-3-319-05005-8

In Chapter 3, please substitute expression (3.6) with the better formulation provided. Please substitute 'wrong' expressions (3.8) and (3.9) with the corresponding 'correct' ones.

(i)   Present Formulation:

$$\forall \omega \in \Omega, \forall \beta \in B, \forall i \in I, \forall h \in C_i$$
$$l_{\beta h i} \geq L_{\omega \beta h i} \vartheta_{\omega i}. \tag{3.6}$$

Better Formulation:

$$\forall \beta \in B, \forall i \in I, \forall h \in C_i$$
$$l_{\beta h i} \geq \sum_{\omega \in \Omega} L_{\omega \beta h i} \vartheta_{\omega i}. \tag{3.6}$$

The online version of the original chapter can be found at
http://dx.doi.org/10.1007/978-3-319-05005-8_3

G. Fasano, *Solving Non-standard Packing Problems by Global Optimization and Heuristics*, SpringerBriefs in Optimization, DOI 10.1007/978-3-319-05005-8, © Giorgio Fasano 2014

(ii) Present Formulation, Wrong:

$$\forall \omega \in \Omega, \forall \beta \in B, \forall i \in I, \forall h \in C_i$$
$$l_{\beta h i} \le L_{\omega \beta h i} \vartheta_{\omega i}. \tag{3.8}$$

Correct Formulation:

$$\forall \beta \in B, \forall i \in I, \forall h \in C_i$$
$$l_{\beta h i} \le \sum_{\omega \in \Omega} L_{\omega \beta h i} \vartheta_{\omega i}. \tag{3.8}$$

(iii) Present Formulation, Wrong:

$$\forall \omega \in \Omega, \forall \beta \in B, \forall i \in I, \forall h \in C_i$$
$$l_{\beta h i} = L_{\omega \beta h i} \vartheta_{\omega i}. \tag{3.9}$$

Correct Formulation:

$$\forall \beta \in B, \forall i \in I, \forall h \in C_i$$
$$l_{\beta h i} = \sum_{\omega \in \Omega} L_{\omega \beta h i} \vartheta_{\omega i}. \tag{3.9}$$

Appendix: Case Studies 1.1–1.20

These cases were solved directly by the general MIP model of Sect. 2.1. Table A.1 reports the overall results.

Table A.2 reports, as an illustrative example, the relevant input data for Case Study 1.5. This instance involves only single parallelepipeds, to be loaded into a domain of units 20, 34, and 50, respectively.

G. Fasano, *Solving Non-standard Packing Problems by Global Optimization and Heuristics*, SpringerBriefs in Optimization, DOI 10.1007/978-3-319-05005-8, © Giorgio Fasano 2014

Table A.1 Case Studies 1.1–1.20 results

Case studies	Total number of single parallelepipeds	Total number of *tetris*-like items	Total number of *components*	Loaded volume % (rounded to nearest)	Centre of mass domain volume % (rounded to the nearest)	CPU time (s)
Case Study 1.1	15	0	15	79.1	0.5	7
Case Study 1.2	20	0	20	80.9	0.5	152
Case Study 1.3	23	0	23	71.4	0.3	20
Case Study 1.4	20	0	20	89.2	0.4	1,394
Case Study 1.5	18	0	18	90.9	0.4	48
Case Study 1.6	25	0	25	87.0	0.3	5,769
Case Study 1.7	5	3	13	90.2	8.7	1
Case Study 1.8	2	6	19	95.6	2.2	1,866
Case Study 1.9	3	4	19	79.3	0.2	2
Case Study 1.10	25	0	25	80.1	0.3	238
Case Study 1.11	0	4	26	70.1	0.4	44
Case Study 1.12	2	6	15	92.9	0.7	419
Case Study 1.13	1	6	15	96.3	1.6	470
Case Study 1.14	2	4	34	66.6	1.2	4
Case Study 1.15	3	5	26	74.0	1.2	59
Case Study 1.16	4	6	18	75.2	0.3	210
Case Study 1.17	16	6	30	78.8	0.5	4,406
Case Study 1.18	3	5	26	69.5	0.2	278
Case Study 1.19	1	4	27	84.2	1.2	2
Case Study 1.20	1	7	28	75.0	5.8	9

Table A.2 Case Study 1.5 item dimensions

Single parallelepiped types	L1 side (units)	L2 side (units)	L3 side (units)
P1	14	20	21
P2	6	12	43
P3	8	13	28
P4	6	21	22
P5	6	20	23
P6	12	13	15
P7	10	12	17
P8	7	12	19
P9	8	14	14
P10	8	10	19
P11	5	7	27
P12	8	9	12
P13	6	10	13
P14	6	6	15
P15	4	11	12
P16	6	6	10
P17	5	6	9
P18	5	5	6

Case Studies 2.2–2.4

Tables A.3, A.4, A.5, and A.6 report the input data relevant to Case Studies 2.2–2.4.

G. Fasano, *Solving Non-standard Packing Problems by Global Optimization and Heuristics*, SpringerBriefs in Optimization, DOI 10.1007/978-3-319-05005-8, © Giorgio Fasano 2014

Table A.3 Case Study 2.2 item dimensions

Single parallelepiped types	Number of single parallelepipeds per type	L1 side (units)	L2 side (units)	L3 side (units)
T1	1	15	30	32
T2	1	10	31	32
T3	1	15	21	22
T4	1	6	10	34
T5	1	6.5	12.5	25
T6	1	4	14	36
T7	8	11	11.43	15
T8	4	10	11.43	15
T9	1	3	20	28
T10	1	5	11.43	15
T11	1	2	12	31
T12	4	3	4	4
T13	4	2	2	10
T14	2	1	2	4

Table A.4 Case Study 2.3 item dimensions

Single parallelepiped types	Number of single parallelepipeds per type	L1 side (units)	L2 side (units)	L3 side (units)
T1	1	1	23	39
T2	1	4	10	20
T3	1	1	23	24
T4	8	10	4	10
T5	1	1	12	18
T6	1	1	10	20
T7	4	4	4	10
T8	1	1	10	14
T9	1	1	6	20
T10	1	1	8	12

Table A.5 Case Study 2.4 item dimensions

Single parallelepiped types	Number of single parallelepipeds per type	L1 side (units)	L2 side (units)	L3 side (units)
T1	1	15	20	25
T2	1	15	15	16
T3	1	15	15	15
T4	8	10	10	25
T5	1	10	15	16
T6	1	10	15	15
T7	4	8	15	15
T8	1	10	10	15
T9	1	7	11	15
T10	1	5	15	15
T11	4	6	8	16

Table A.6 Case Study 2.2–2.4 domain dimensions

Case studies	Domain side (units)	Domain side (units)	Domain side (units)
Case Study 2.2	39	48.3	47.5
Case Study 2.3	39	73.2	47.5
Case Study 2.4	26	38	42

Case Studies 3.2 and 3.3

Tables A.7 and A.8 report the item input data relevant to Case Studies 3.2 and 3.3. Table A.9 reports the relevant domain dimensions.

G. Fasano, *Solving Non-standard Packing Problems by Global Optimization and Heuristics*, SpringerBriefs in Optimization, DOI 10.1007/978-3-319-05005-8, © Giorgio Fasano 2014

Table A.7 Case Study 3.2 item dimensions

Single parallelepiped types	Number of single parallelepipeds per type	L1 side (units)	L2 side (units)	L3 side (units)
T1	1	5	39	49.4
T2	1	12	19	34.4
T3	1	10	22	34
T4	1	5	24	39.4
T5	1	5	29	32
T6	1	5	24	34.4
T7	1	5	22	34
T8	1	5	20	34.4
T9	1	5	19	34.4
T10	1	5	18.9	34.4
T11	3	5	20	20
T12	1	5	14	39.4
T13	2	5	10	54.4
T14	1	5	22	24
T15	2	5	15	34.4
T16	1	5	15	34
T17	1	5	10	49.4
T18	1	5	12	39.4
T19	3	5	10	44.4
T20	1	5	17	24
T21	1	10	10	20
T22	1	5	20	20
T23	1	5	10	39.4
T24	1	5	10	39
T25	1	5	7	54.4
T26	1	5	10	34.4
T27	1	5	15	22
T28	3	5	10	32
T29	1	5	10	29
T30	1	5	7	39.4
T31	1	5	5	54.4
T32	2	5	7	34.4
T33	1	5	10	24
T34	3	5	10	22
T35	1	5	5	42
T36	1	10	10	10
T37	1	5	10	20
T38	1	5	5	39.4
T39	1	5	5	39
T40	1	5	5	37

Table A.8 Case Study 3.3 item dimensions

Single parallelepiped types	Number of single parallelepipeds per type	L1 side (units)	L2 side (units)	L3 side (units)
T1	1	19	33	45
T2	1	21	21	50
T3	1	6	33	45
T4	4	10	11.43	15
T5	33	5	6	9
T6	33	5	11.43	15
T7	2	5	9	9
T8	2	5	5	9
T9	4	5	5	5
T10	1	12	16.51	16.87
T11	2	2	5	5

Table A.9 Case Studies 3.2 and 3.3 domain dimensions

Case studies	Domain side (units)	Domain side (units)	Domain side (units)
Case Study 3.2	74.4	49	42
Case Study 3.3	39	73.2	47.5

Case Study 6

Table A.10 summarizes the process carried out to solve Case Study 6.

G. Fasano, *Solving Non-standard Packing Problems by Global Optimization and Heuristics*, SpringerBriefs in Optimization, DOI 10.1007/978-3-319-05005-8, © Giorgio Fasano 2014

Table A.10 Case Study 6—solution process

Main cycle	Step	No. of steps	No. of selected items	No. of loaded items	Added items	Occupied volume (%)	CPU time (s)
Basic Cycle 1	*Initialization*	1	20	0	0	0.00	00:00:02
	Packing	1	20	16	16	26.16	00:00:03
	Item-exchange	0	20	16	0	26.16	00:00:00
	Hole-filling	2	27	27	11	34.40	00:00:03
Basic Cycle 2	*Initialization*	1	44	27	0	34.40	00:00:08
	Packing	1	44	34	7	39.04	00:00:02
	Item-exchange	1	44	33	−1	38.36	00:00:01
	Hole-filling	3	56	56	23	49.78	00:00:16
Basic Cycle 3	*Initialization*	1	68	56	0	49.78	00:00:08
	Packing	1	68	60	4	51.41	00:00:02
	Item-exchange	0	68	60	0	51.41	00:00:00
	Hole-filling	30	81	81	21	59.82	00:01:00
Basic Cycle 4	*Initialization*	1	86	81	0	59.82	00:00:07
	Packing	1	86	81	0	59.82	00:00:02
	Item-exchange	0	86	81	0	59.82	00:00:00
	Hole-filling	30	96	96	15	64.73	00:01:13
Basic Cycle 5	*Initialization*	1	107	96	0	64.73	00:00:17
	Packing	1	107	98	2	65.33	00:00:03
	Item-exchange	0	107	98	0	65.33	00:00:00
	Hole-filling	34	115	115	17	69.85	00:02:14
Basic Cycle 6	*Initialization*	1	129	115	0	69.85	00:00:40
	Packing	1	129	118	3	70.51	00:00:05
	Item-exchange	1	129	118	0	70.52	00:00:05
	Hole-filling	40	138	138	20	74.85	00:04:09
Basic Cycle 7	*Hole-filling*	117	255	255	117	85.77	00:38:00
	Item-exchange	0	255	255	0	85.77	00:00:00

Virtual Items

Outcomes relevant to a set of 32 *virtual* item tests are reported in Table A.11.

The tests were executed by using the IBM ILOG CPLEX Optimizer 12.3 (IBM corporation 2010) on a personal computer (Core 2 Duo P8600, 2.40 GHz processor; 1.93 GB RAM; MS Windows XP Professional, Service Pack 2).

G. Fasano, *Solving Non-standard Packing Problems by Global Optimization and Heuristics*, SpringerBriefs in Optimization, DOI 10.1007/978-3-319-05005-8, © Giorgio Fasano 2014

Table A.11 *Virtual* item case studies

Case studies	Total no. of tetris-like items	Total no. of *components*	Already loaded volume % (rounded to nearest)	Total no. of *virtual* items	Tot. *virtual* item v olume % (rounded to nearest)	Residual-free volume % (rounded to nearest)	CPU time (s)
Case Study 1	6	11	76.8	5	18.4	4.8	92
Case Study 2	6	9	45.5	10	30.2	24.3	256
Case Study 3	8	13	69.6	10	23.3	5.3	188
Case Study 4	4	34	36.8	10	45.6	17.5	319
Case Study 5	7	15	56.9	10	27.7	15.3	247
Case Study 6	6	16	49.8	10	33.2	17.1	281
Case Study 7	5	37	29.0	10	49.3	21.7	242
Case Study 8	5	44	30.0	10	57.1	13.0	203
Case Study 9	4	32	51.4	10	22.4	26.0	323
Case Study 10	5	19	39.2	10	33.8	27.0	242
Case Study 11	7	12	63.7	1	0.4	36.0	24
Case Study 12	4	29	44.1	10	38.1	18.0	333
Case Study 13	6	12	29.7	10	38.8	32.0	213

Case Study 14	5	33	49.9	10	12.7	37.3	198
Case Study 15	3	28	47.6	10	26.8	25.7	189
Case Study 16	5	23	48.2	10	12.5	39.2	206
Case Study 17	3	27	33.4	10	25.9	40.7	2,300
Case Study 18	4	27	33.5	10	35.6	30.9	261
Case Study 19	3	30	33.8	10	17.3	48.9	378
Case Study 20	9	49	46.7	10	39.2	14.2	212
Case Study 21	8	31	56.7	10	24.9	18.4	313
Case Study 22	5	33	43.8	10	34.0	22.2	239
Case Study 23	5	63	42.1	10	27.8	30.1	225
Case Study 24	5	51	45.3	10	32.3	22.4	214
Case Study 25	6	47	40.3	10	32.5	27.2	217
Case Study 26	4	21	46.5	8	15.1	38.4	284
Case Study 27	6	27	41.1	10	22.0	37.0	339
Case Study 28	5	26	49.9	10	10.0	40.2	251
Case Study 29	5	22	42.0	10	17.0	41.0	227
Case Study 30	5	25	42.9	10	26.8	30.4	219
Case Study 31	22	54	54.4	10	24.0	21.7	351
Case Study 32	3	15	47.93	10	35.9	20.0	258

References

Aardal, K., Pochet, Y., Wolsey, L.A.: Capacitated facility location: valid inequalities and facets. Math. Oper. Res. **20**, 562–582 (1995)

Addis, B., Locatelli, M., Schoen, F.: Efficiently packing unequal disks in a circle: a computational approach which exploits the continuous and combinatorial structure of the problem. Oper. Res. Lett. **36**(1), 37–42 (2008a)

Addis, B., Locatelli, M., Schoen, F.: Disk packing in a square: a new global optimization approach. INFORMS J. Comput. **20**(4), 516–524 (2008b)

Allen, S.D., Burke, E.K., Kendall, G.: A hybrid placement strategy for the three-dimensional strip packing problem. Eur. J. Oper. Res. **209**(3), 219–227 (2011)

Allen, S.D., Burke, E.K., Mareček, J.: A space-indexed formulation of packing boxes into a larger box. Oper. Res. Lett. **40**(1), 20–24 (2012)

Alvim, A.C., Ribeiro, C.C.: A hybrid bin-packing heuristic to multiprocessor scheduling. In: Ribeiro, C.C., Martins, S.L. (eds.) Lecture Notes in Computer Science, vol. 3059, pp. 1–13. Springer, Berlin (2004)

Andersen, K., Cornuéjols, G., Li, Y.: Reduce-and-split cuts: improving the performance of mixed integer Gomory cuts. Manag. Sci. **51**, 1720–1732 (2005)

Andreello, G., Caprara, A., Fischetti, M.: Embedding cuts in a branch and cut framework: a computational study with {0, 1/2}-cuts. INFORMS J. Comput. **19**, 229–238 (2007)

Atamtürk, A.: Cover and pack inequalities for (mixed) integer programming. Ann. Oper. Res. **139**, 21–38 (2005)

Ausiello, G., Crescenzi, P., Gambosi, G., Kann, V., Marchetti-Spaccamela, A., Protasi, M.: Complexity and Approximation (Corrected edn.). Springer, Berlin (2003). ISBN 978-3540654315

Balas, E., Ceria, S., Cornuéjols, G.: Mixed 0-1 programming by lift-and-project in a branch-and-cut framework. Manag. Sci. **42**, 1229–1246 (1996)

Beasley, J.E.: An exact two-dimensional non-guillotine cutting tree search procedure. Oper. Res. **33**(1), 49–64 (1985)

Bennell, J.A., Han, W., Zhao, X., Song, X.: Construction heuristics for two-dimensional irregular shape bin packing with guillotine constraints. Eur. J. Oper. Res. **230**(3), 495–504 (2013)

Bennel, J.A., Lee, L.S., Potts, C.N.: A genetic algorithm for two-dimensional bin packing with due dates. Int. J. Prod. Econ. **145**(2), 547–560 (2013)

Bennell, J., Oliveira, J.: The geometry of nesting problems: a tutorial. Eur. J. Oper. Res. **184**, 397–415 (2008)

Bennell, J.A., Oliveira, J.F.: A tutorial in irregular shape packing problems. J. Oper. Res. Soc. **60**(S1), S93–S105 (2009) de Berg, M., van Kreveld, M.J., Overmars, M., Schwarzkopf, O.: Polygon triangulation. In: de Berg, M., van Kreveld, M.J., Overmars, M., Schwarzkopf, O. (eds.) Computational Geometry, pp. 45–61. Springer, Berlin (2000). ISBN 3-540-65620

Betke, U., Henk, M.: Densest lattice packings of 3-polytopes. Comput. Geom. **16**(3), 157–186 (2000)

Birgin, E.G., Lobato, R.D.: Orthogonal packing of identical rectangles within isotropic convex regions. Comput. Ind. Eng. **59**(4), 595–602 (2010)

Birgin, E., Martinez, J., Nishihara, F.H., Ronconi, D.P.: Orthogonal packing of rectangular items within arbitrary convex regions by nonlinear optimization. Comput. Oper. Res. **33**(12), 3535–3548 (2006)

Bischoff, E.E.: Three-dimensional packing of items with limited load bearing strength. Eur. J. Oper. Res. **168**(3), 952–966 (2006)

Bischoff, E.E., Ratcliff, M.S.W.: Issues in the development of approaches to container loading. OMEGA **23**(4), 377–390 (1995)

Bortfeldt, A., Gehring, H.: A hybrid genetic algorithm for the container loading problem. Eur. J. Oper. Res. **131**(1), 143–161 (2001)

Bortfeldt, A., Gehring, H., Mack, D.: A parallel tabu search algorithm for solving the container loading problem. Parallel Comput. **29**(5), 641–662 (2003)

Bortfeldt, A., Wäscher, G.: Container loading problems—a state-of-the-art review. FEMM working papers 120007, Otto-von-Guericke University Magdeburg, Faculty of Economics and Management, Magdeburg (2012)

Burke, E.K., Hellier, R., Kendall, G., Whitwell, G.: A new bottom-left-fill heuristic algorithm for the 2D irregular packing problem. Oper. Res. **54**(3), 587–601 (2006)

Burke, E.K., Guo, Q., Hellier, R., Kendall, G.: A hyper-heuristic approach to strip packing problems. In: Proceedings of the 11th International Conference on Parallel Problem Solving from Nature: Part I, pp. 465–474. Springer, Berlin (2010)

Cagan, J., Shimada, K., Yin, S.: A survey of computational approaches to three-dimensional layout problems. Comput. Aided Des. **34**, 597–611 (2002)

Caprara, A., Monaci, M.: On the 2-dimensional knapsack problem. Oper. Res. Lett. **1**(32), 5–14 (2004)

Cassioli, A., Locatelli, M.: A heuristic approach for packing identical rectangles in convex regions. Comput. Oper. Res. **38**(9), 1342–1350 (2011)

Castillo, I., Kampas, F.J., Pintér, J.D.: Solving circle packing problems by global optimization: numerical results and industrial applications. Eur. J. Oper. Res. **191**(3), 786–802 (2008)

Ceria, S., Cordier, C., Marchand, H., Wolsey, L.A.: Cutting planes for integer programs with general integer variables. Math. Program. **81**, 201–214 (1998)

Chen, C.S., Lee, S.M., Shen, Q.S.: An analytical model for the container loading problem. Eur. J. Oper. Res. **80**, 68–76 (1995)

Chen, D.S., Batson, R.G., Dang, Y.: Applied Integer Programming—Modeling and Solutions. Wiley, Hoboken, NJ (2010)

Chernov, N., Stoyan, Y.G., Romanova, T.: Mathematical model and efficient algorithms for object packing problem. Comput. Geom. Theor. Appl. **43**(5), 535–553 (2010)

Chernov, N., Stoyan, Y., Romanova, T., Pankratov, A.: Phi-functions for 2D objects formed by line segments and circular arcs. Adv. Oper. Res. (2012). doi:10.1155/2012/346358

Chiong, R., Dhakal, S. (eds.): Natural Intelligence for Scheduling, Planning and Packing Problems. Springer, Berlin (2009)

Chlebík, M., Chlebíková, J.: Inapproximability results for orthogonal rectangle packing problems with rotations. In: Proceedings of 6th Conference on Algorithms and Complexity (CIAC 2006), LNCS, pp. 199–210. Springer, Rome, Italy (2006)

Christensen, S.G., Rousøe, D.M.: Container loading with multi-drop constraints. Int. Trans. Oper. Res. **16**(6), 727–743 (2009)

Coffman, E., Garey, J.M., Johnson, D.: Approximation Algorithms for Bin Packing: A Survey. PWS Publishing Company, Boston (1997)

Constantino, M.: Lower bounds in lot-sizing models: a polyhedral study. Math. Oper. Res. **23**, 101–118 (1998)

Cordier, C., Marchand, H., Laundry, R., Wolsey, L.A.: A branch-and-cut code for mixed integer programs. Math. Program. **86**, 335–354 (2001)

Cornuéjols, G.: Valid inequalities for mixed integer linear programs. Math. Program. **112**, 3–44 (2008)

Dash, S., Günlük, O., Lodi, A.: MIR closures of polyhedral sets. Math. Program. **121**, 33–60 (2010)

De Farias, I.R., Johnson, E.L., Nemhauser, G.L.: Facets of the complementarity knapsack polytope. Technical Report LEC-98-08, Georgia Institute of Technology, Atlanta (1998)

De Loera, J., Hemmecke, R., Köppe, M.: Global mixed-integer polynomial optimization via summation. In: De Loera, J., Hemmecke, R., Köppe, M. (eds.) Algebraic and Geometric Ideas in the Theory of Discrete Optimization. MPS-SIAM Series on Optimization, pp. 157–177. Society for Industrial and Applied Mathematics, Philadelphia (2012)

Dowsland, K.A., Herbert, E.A., Kendall, G., Burke, E.: Using tree search bounds to enhance a genetic algorithm approach to two rectangle packing problems. Eur. J. Oper. Res. **168**(2), 390–402 (2006)

Dyckhoff, H., Scheithauer, G., Terno, J.: Cutting and packing. In: Dell'Amico, M., Maffioli, F., Martello, S. (eds.) Annotated Bibliographies in Combinatorial Optimization, pp. 393–412. Wiley, Chichester (1997)

Egeblad, J.: Placement of two- and three-dimensional irregular shapes for inertia moment and balance. In: Morabito, R., Arenales, M.N., Yanasse, H.H. (eds.) Int. Trans. Oper. Res. (Special Issue on Cutting, Packing and Related Problems) **16**(6), 789–807 (2009)

Egeblad, J., Nielsen, B.K., Brazil, M.: Translational packing of arbitrary polytopes. Comput. Geom. **42**(4), 269–288 (2009b)

Egeblad, J., Nielsen, B.K., Odgaard, A.: Fast neighborhood search for two-and three-dimensional nesting problems. Eur. J. Oper. Res. **183**(3), 1249–1266 (2007)

Egeblad, J., Pisinger, D.: Heuristic approaches for the two- and three-dimensional knapsack packing problems. DIKU Technical-Report No. 2006-13, SSN: 0107-8283, Department of Computer Science, University of Copenhagen, Denmark (2006)

Egeblad, J., Pisinger, D.: Heuristic approaches for the two- and three-dimensional knapsack packing problem. Comput. Oper. Res. **36**, 1026–1049 (2009)

Eley, M.: Solving container loading problems by block arrangement. Eur. J. Oper. Res. **141**(2), 393–409 (2002)

Escudero, L.: S3 sets. An extension of the Beale-Tomlin special ordered sets. Math. Program. **42**, 113–123 (1988)

Faroe, O., Pisinger, D., Zachariasen, M.: Guided local search for the three-dimensional bin packing problem. INFORMS J. Comput. **15**(3), 267–283 (2003)

Fasano, G.: Satellite Optimal Layout. Application of Mathematical and Optimization Techniques. IBM Europe Institute, Garmisch-Partenkirchen, Germany (1989)

Fasano, G.: Cargo analytical integration in space engineering: a three-dimensional packing model. In: Ciriani, T., Gliozzi, S., Johnson, E.L. (eds.) Operations Research in Industry, pp. 232–246. Macmillan, London (1999)

Fasano, G.: MIP models for solving 3-dimensional packing problems arising in space engineering. In: Ciriani, T., Fasano, G., Gliozzi, S., Tadei, R. (eds.) Operations Research in Space and Air, pp. 43–56. Kluwer Academic Publisher, Boston (2003)

Fasano, G.: A MIP approach for some practical packing problems: balancing constraints and tetris-like items. 4OR Q. J. Oper. Res. **2**(2), 161–174 (2004)

Fasano, G.: MIP-based heuristic for non-standard 3D-packing problems. 4OR Q. J. Oper. Res. **6**(3), 291–310 (2008)

Fasano, G.: A multi-level MIP-based heuristic approach for the cargo accommodation of a space vehicle. In: 6th ESICUP Meeting, Valencia, Spain, 25–29 Mar 2009

Fasano, G.: A global optimization point of view to handle non-standard objective packing problems. J. Global. Optim. **55**(2), 279–299 (2013)

Fasano, G., Castellazzo, A.: Approximated solutions to a 3D-packing MIP model by a non-linear approach. Comm. Appl. Ind. Math. (2013). doi: 10.1685/journal.caim.449

Fasano, G., Vola, M.C.: Space module on-board stowage optimization exploiting containers' empty volumes. In: Fasano, G., Pintér, J.D. (eds.) Modeling and Optimization in Space Engineering, pp. 249–269. Springer Science + Business Media, New York (2013)

Fekete, S.P., Schepers, J.: A new exact algorithm for general orthogonal d-dimensional knapsack problems. In: Burkard, R., Woeginger, G. (eds.) Algorithms ESA '97. Springer Lecture Notes in Computer Science, vol. 1284, pp. 144–156. Springer, Berlin (1997)

Fekete, S., Schepers, J.: A combinatorial characterization of higher-dimensional orthogonal packing. Math. Oper. Res. **29**, 353–368 (2004)

Fekete, S., Schepers, J., van der Veen, J.C.: An exact algorithm for higher-dimensional orthogonal packing. Oper. Res. **55**(3), 569–587 (2007)

Fischetti, M., Luzzi, I.: Mixed-integer programming models for nesting problems. J. Heuristics **15**(3), 201–226 (2009)

Floudas, C.A., Akrotirianakis, I.G., Caratzoulas, S., Meyer, C.A., Kallrath, J.: Global optimization in the 21st century: advances and challenges for problems with nonlinear dynamics. Comp. Chem. Eng. **29**, 1185–1202 (2005)

Floudas, C.A., Pardalos, P.M.: A Collection of Test Problems for Constrained Global Optimization Algorithms. Springer, New York (1990)

Floudas, C.A., Pardalos, P.M., et al.: Handbook of Test Problems in Local and Global Optimization, Nonconvex Optimization and Its Applications Series 33. Kluwer Academic Publishers, Dordrecht, The Netherlands (1999)

Floudas, C.A., Pardalos, P.M. (eds.): Encyclopedia of Optimization. Kluwer Academic Publishers, Dordrecht, The Netherlands (2001)

Gan, M., Gopinathan, N., Jia, X., Williams, R.A.: Predicting packing characteristics of particles of arbitrary shapes. KONA **22**, 82–93 (2004)

Gehring, H., Bortfeldt, A.: A parallel genetic algorithm for solving the container loading problem. Int. Trans. Oper. Res. **9**(4), 497–511 (2002)

Gensane, T.: Dense packing of equal spheres in a cube. Electron. J. Combinator. **11**(1), 1–17 (2004)

Goldreich, O.: Computational Complexity: A Conceptual Perspective. Cambridge University Press, Cambridge (2008)

Golomb, S.W.: Polyominoes, 2nd edn. Princeton University Press, Princeton, NJ (1994). ISBN 0-691-02444-8

Gomes, A.M., Olivera, J.F.: A 2-exchange heuristics for nesting problems. Eur. J. Oper. Res. **141**, 359–570 (2002)

Gonçalves, J.F., Resende, M.G.: A parallel multi-population biased random-key genetic algorithm for a container loading problem. Comput. Oper. Res. **39**(2), 179–190 (2012)

Gray, J.J.: The Hilbert Challenge. Oxford University Press, Oxford (2000)

Hadjiconstantinou, E., Christofides, N.: An exact algorithm for general, orthogonal, two-dimensional knapsack problems. Eur. J. Oper. Res. **83**(1), 39–56 (1995)

Hamacher, H.W., Labbé, M., Nickel, S., Sonneborn, T.: Adapting polyhedral properties from facility to hub location problems. Discrete. Appl. Math. **145**, 104–116 (2004)

Hanzon, B., Jibetean, D.: Global minimization of a multivariate polynomial using matrix methods. J. Global. Optim. **27**, 1–23 (2003)

Hifi, M., M'Hallah, R.: A literature review on circle and sphere packing problems: models and methodologies. Adv. Oper. Res. Article ID 150624. http://downloads.hindawi.com/journals/aor/2009/150624.pdf (2009). Accessed 30 Aug 2013

Hopper, E., Turton, B.C.: A review of the application of meta-heuristic algorithms to 2D strip packing problems. Artif. Intell. Rev. **16**(4), 257–300 (2001)

Hopper, E., Turton, B.C.: An empirical study of meta-heuristics applied to 2D rectangular bin packing—part II. Studia Informatica Universalis **2**(1), 93–106 (2002)

Horst, R., Pardalos, P.M. (eds.): Handbook of Global Optimization. Kluwer Academic Publishers, Dordrecht, The Netherlands (1995)

Horst, R., Tuy, H.: Global Optimization: Deterministic Approaches, 3rd edn. Springer, Berlin (1996)

Horst, R., Pardalos, P.M. (eds.): Developments in Global Optimization. Kluwer Academic Publishers, Dordrecht, The Netherlands (1997)

Ibaraki, T., Imahori, S., Yagiura, M.: Hybrid metaheuristics for packing problems. In: Blum, C., Aguilera, M.J., Roli, A., Sampels, M. (eds.) Hybrid Metaheuristics: An Emerging Approach to Optimization. Studies in Computational Intelligence (SCI), vol. 114, pp. 185–219. Springer, Berlin (2008)

IBM Corporation: ILOG CPLEX Optimizer. High Performance Mathematical Optimization Engines. IBM Corporation Software Group, Somers, NY, USA (2010). WSD14044-USEN-01

INFORMS Computing Society: Mathematical Programming Glossary. http://glossary.computing. society.informs.org (2010). Accessed 30 Aug 2013

Iori, M., Martello, S., Monaci, M.: Metaheuristic algorithms for the strip packing problem. In: Pardalos, P.M., Korotkikh, V. (eds.) Optimization and Industry: New Frontiers, pp. 159–179. Kluwer Academic Publishers, Dordrecht, The Netherlands (2003)

Jünger, M., Liebling, T.M., Naddef, D., Nemhauser, G.L., Pulleyblank, W.R., et al. (eds.): 50 Years of Integer Programming 1958-2008: From the Early Years to the State-of-the-Art. Springer, Berlin (2009). ISBN 978-3-540-68247-5

Junqueira, L., Morabito, R., Yamashita, D.S.: Mip-based approaches for the container loading problem with multi-drop constraints. Ann. Oper. Res. **199**(1), 51–75 (2011)

Junqueira, L., Morabito, R., Yamashita, D.S.: Three-dimensional container loading models with cargo stability and load bearing constraints. Comput. Oper. Res. **39**(1), 74–85 (2012)

Junqueira, L., Morabito, R., Yamashita, D.S., Yanasse, H.H.: Optimization models for the three-dimensional container loading problem with practical constraints. In: Fasano, G., Pintér, J.D. (eds.) Modeling and Optimization in Space Engineering, pp. 271–294. Springer Science + Business Media, New York (2013)

Kahng, A.B., Wang, Q.: Implementation and extensibility of an analytic placer. IEEE Trans. Comput. Aided Des. Integrated Circ. Syst. **24**(5), 734–747 (2005)

Kallrath, J.: Mixed-integer nonlinear applications. In: Ciriani, T., Ghiozzi, S., Johnson, E.L. (eds.) Operations Research in Industry, pp. 42–76. Macmillan, London (1999)

Kallrath, J.: Modeling difficult optimization problems. In: Floudas, C.A., Pardalos, P.M. (eds.) Encyclopedia of Optimization, 2nd edn, pp. 2284–2297. Springer, New York (2008)

Kallrath, J.: Cutting circles and polygons from area-minimizing rectangles. J. Global. Optim. **43** (2–3), 299–328 (2009)

Kampas, F.J., Pintér, J.D.: Packing Equal-Size Circles in a Triangle. http://library.wolfram.com/ infocenter/TechNotes/6202 (2005). Accessed 30 Aug 2013

Kampas, F.J., Pintér, J.D.: Optimization with Mathematica: Scientific, Engineering, and Economic Applications. Springer Science + Business Media, New York (forthcoming)

Kang, M.K., Jang, C.S., Yoon, K.S.: Heuristics with a new block strategy for the single and multiple container loading problems. J. Oper. Res. Soc. **61**, 95–107 (2010)

Kenmochi, M., Imamichi, T., Nonobe, K., Yagiura, M., Nagamochi, H.: Exact algorithms for the two-dimensional strip packing problem with and without rotations. Eur. J. Oper. Res. **198**, 73–83 (2009)

Kim, J.G., Kim, Y.D.: A linear programming-based algorithm for floorplanning in VLSI design. IEEE Trans. Comput. Aided Des. Integrated Circ. Syst. **22**(5), 584–592 (2003)

Kleinhans, J.M., Sigl, G., Johannes, F.M., Antreich, K.J.: GORDIAN: VLSI placement by quadratic programming and slicing optimization. IEEE Trans. Comput. Aided Des. Integrated Circ. Syst. **10**(3), 356–365 (1991)

Lai, K.K., Xue, J., Xu, B.: Container packing in a multi-customer delivering operation. Comput. Ind. Eng. **35**(1–2), 323–326 (1998)

Li, H.L., Chang, C.T., Tsai, J.F.: Approximately global optimization for assortment problems using piecewise linearization techniques. Eur. J. Oper. Res. **140**, 584–589 (2002)

Liberti, L., Maculan, N. (eds.): Global Optimization: From Theory to Implementation. Springer Science + Business Media, New York (2005)

Limbourg, S., Schyns, M., Laporte, G.: Automatic aircraft cargo load planning. J. Oper. Res. Soc. **63**, 1271–1283 (2012)

Linderoth, J.T., Savelsbergh, M.W.: A computational study of search strategies for mixed integer programming. INFORMS J. Comput. **11**(2), 173–187 (1999)

Liu, C., Baskiyar, S.: Scheduling mixed tasks with deadlines in grids using bin packing. In: International Conference on Parallel and Distributed Systems—ICPADS (2008). doi:10.1109/ICPADS.2008.127

Locatelli, M., Raber, U.: Packing equal circles into a square: a deterministic global optimization approach. Discrete Appl. Math. **122**, 139–166 (2002)

Lodi, A., Martello, S., Monaci, M., Vigo, D.: Two-dimensional bin packing problems. In: Paschos, V.T. (ed.) Paradigms of Combinatorial Optimization, pp. 107–129. Wiley/ISTE, Hoboken, NJ (2010)

López-Camacho, E., Ochoa, G., Terashima-Marín, H., Burke, E.: An effective heuristic for the two-dimensional irregular bin packing problem. Ann. Oper. Res. **206**(1), 241–264 (2013)

Mack, D., Bortfeldt, A., Gehring, H.: A parallel hybrid local search algorithm for the container loading problem. Int. Trans. Oper. Res. **11**(5), 511–533 (2004)

Marchand, H., Martin, A., Weismantel, R., Wolsey, L.A.: Cutting planes in integer and mixed integer programming. Technical Report CORE DP9953, Université Catholique de Louvain, Louvain-la-Neuve, Belgium (1999)

Marchand, H., Wolsey L.A.: Aggregation and mixed integer rounding to solve MIPs. Technical Report CORE DP9839, Université Catholique de Louvain, Louvain-la-Neuve, Belgium (1998)

Martello, S., Pisinger, D., Vigo, D.: The three-dimensional bin packing problem. Oper. Res. **48**(2), 256–267 (2000)

Martello, S., Pisinger, D., Vigo, D., Den Boef, E., Korst, J.: Algorithms for general and robot-packable variants of the three-dimensional bin packing problem. ACM Trans. Math. Software **33**(1), 7 (2007)

Minoux, M., Vajda, S.: Mathematical Programming: Theory and Algorithms. Wiley, London (1986)

Morabito, R., Arenales, M.: An and/or-graph approach to the container loading problem. Int. Trans. Oper. Res. **1**(1), 59–73 (1994)

Moura, A., Oliveira, J.F.: A GRASP approach to the container-loading problem. IEEE Intell. Syst. **4**(20), 50–57 (2005)

Nemhauser, G.L., Wolsey, L.A.: Integer and Combinatorial Optimization. Wiley, New York (1988)

Nemhauser, G.L., Wolsey, L.A.: A recursive procedure for generating all cuts for 0-1 mixed integer programs. Math. Program. **46**, 379–390 (1990)

Oliveira, J.F., Gomes, A.M., Ferreira, J.S.: TOPOS—a new constructive algorithm for nesting problems. OR Spectrum **22**(2), 263–284 (2000)

Padberg, M.W.: Linear Optimization and Extensions. Springer, Heidelberg (1995)

Padberg, M.W.: Packing small boxes into a big box. Office of Naval Research, N00014-327, New York University (1999)

Padberg, M.W.: Classical cuts for mixed integer programming and branch-and-cut. Math. Meth. Oper. Res. **53**, 173–203 (2001)

Padberg, M.W., Rinaldi, G.: A branch and cut algorithm for the resolution of large-scale symmetric traveling salesmen problems. SIAM Rev. **33**, 60–100 (1991)

Padberg, M.W., Van Roy, T.J., Wolsey, L.A.: Valid inequalities for fixed charge problems. Oper. Res. **33**, 842–861 (1985)

Pan, P., Liu, C.L.: Area minimization for floorplans. IEEE Trans. Comput. Aided Des. Integrated Circ. Syst. **14**(1), 123–132 (2006). http://ieeexplore.ieee.org/xpl/RecentIssue.jsp?punumber=43

Pan, Y., Shi, L.: On the equivalence of the max-min transportation lower bound and the time-indexed lower bound for single-machine scheduling problems. Math. Programm Ser. A **110**(3), 543–559 (2007)

Pardalos, P.M., Resende, M.G. (eds.): Handbook of Applied Optimization. Oxford University Press, Oxford (2002)

Pardalos, P.M., Romeijn, H.E. (eds.): Handbook of Global Optimization. Kluwer Academic Publishers, Dordrecht, The Netherlands (2002)

Parreño, F., Alvarez-Valdes, R., Oliveira, J.F., Tamarit, J.M.: A maximal-space algorithm for the container loading problem. INFORMS J. Comput. **20**(3), 412–422 (2008). http://dl.acm.org/citation.cfm?id=1528512

Pesch, E.: Fast Truck-Packing of 3D Boxes. Working paper

Pintér, J.D.: Global Optimization in Action. Kluwer Academic Publishers, Dordrecht, The Netherlands (1996)

Pintér, J.D.: LGO—a program system for continuous and Lipschitz optimization. In: Bomze, I.M., Csendes, T., Horst, R., Pardalos, P.M. (eds.) Developments in Global Optimization, pp. 183–197. Kluwer Academic Publishers, Dordrecht, The Netherlands (1997)

Pintér, J.D.: Global optimization: software, test problems, and applications. In: Pardalos, P.M., Romeijn, H.E. (eds.) Handbook of Global Optimization, vol. 2, pp. 515–569. Kluwer Academic Publishers, Dordrecht, The Netherlands (2002)

Pintér, J.D.: Nonlinear optimization in modeling environments: software implementations for compilers, spreadsheets, modeling languages, and integrated computing systems. In: Jeyakumar, V., Rubinov, A.M. (eds.) Continuous Optimization: Current Trends and Modern Applications, pp. 147–173. Springer Science + Business Media, New York (2005)

Pintér, J.D.: Nonlinear optimization with GAMS /LGO. J. Global Optim. **38**, 79–101 (2007)

Pintér, J.D.: Software development for global optimization. In: Pardalos, P.M., Coleman, T.F. (eds.) Global Optimization: Methods and Applications, Fields Institute Communications, vol. 55, pp. 183–204. American Mathematical Society, Providence, RI (2009)

Pintér, J.D.: Mathematical programming glossary supplement: global optimization. In: Mathematical Programming Glossary. INFORMS Computing Society. http://glossary.computing.society.informs.org (2006). Accessed 30 Aug 2013

Pintér Consulting Services: LGO—A Model Development and Solver System for Global-Local Nonlinear Optimization. User's Guide. Pintér Consulting Services, Inc., Halifax, Canada. www.pinterconsulting.com (2013)

Pisinger, D.: Heuristics for the container loading problem. Eur. J. Oper. Res. **141**(2), 382–392 (2002)

Pisinger, D., Sigurd, M.: The two-dimensional bin packing problem with variable bin sizes and costs. Discrete Optim. **2**(2), 154–167 (2005)

Pisinger, D., Sigurd, M.: Using decomposition techniques and constraint programming for solving the two-dimensional bin packing problem. INFORMS J. Comput. **19**(1), 36–51 (2007)

Pochet, Y., Wolsey, L.A.: Polyhedra for lot-sizing with Wagner-Whitin costs. Math. Program. **67**, 297–324 (1994)

Ramakrishnan, K., Bennel, J.A., Omar, M.K.: Solving two dimensional layout optimization problems with irregular shapes by using meta-heuristic. In: 2008 IEEE International Conference on Industrial Engineering and Engineering Management, pp. 178–182 (2009). doi:10.1109/IEEM.2008.4737855

Ratcliff, M.S.W., Bischoff, E.E.: Allowing for weight considerations in container loading. OR Spektrum **20**(1), 65–71 (1998)

Rebennack, S., Kallrath, J., Pardalos, P.M.: Column enumeration based decomposition techniques for a class of non-convex MINLP problems. J. Global Optim. **43**(2–3), 277–297 (2009)

Scheithauer, G., Stoyan, Y.G., Romanova, T.Y.: Mathematical modeling of interactions of primary geometric 3D objects. Cybern. Syst. Anal. **41**, 332–342 (2005)

Schweighofer, M.: Global optimization of polynomials using gradient tentacles and sums of squares. SIAM J. Optim. **17**(3), 920–942 (2006)

Silva, J.L.C., Soma, N.Y., Maculan, N.: A greedy search for the three-dimensional bin packing problem: the packing static stability case. Int. Trans. Oper. Res. **10**(2), 141–153 (2003)

Specht, E. http://www.packomania.com (2012). Accessed 30 Aug 2013

Stoyan, Y.G., Chugay, A.M.: Packing cylinders and rectangular cuboids with distances between them into a given region. Eur. J. Oper. Res. **197**, 446–455 (2009)

Stoyan, Y.G., Novozhilova, M.V., Kartashov, A.V.: Mathematical model and method of searching for a local extremum for non-convex oriented polygons allocation problem. Eur. J. Oper. Res. **92**, 193–210 (1996)

Stoyan, Y., Scheithauer, G., Gil, N., Romanova, T.: Φ-functions for complex 2D-objects. 4OR Q. J. Belgian French Italian Oper. Res. Soc. **2**(1), 69–84 (2004)

Stoyan, Y., Terno, J., Scheithauer, G., Gil, M., Romanova, T.: Construction of a Phi-function for two convex polytopes. Appl. Math. **29**(2), 199–218 (2002)

Stoyan, Y.G., Yaskov, G.: Packing identical spheres into a rectangular parallelepiped. In: Bortfeldt, A., Homberger, J., Kopfer, H., Pankratz, G., Strangmeier, R. (eds.) Intelligent Decision Support. Current Challenges and Approaches, pp. 47–67. GWV Fachverlage GMbH, Wiesbaden, Germany (2008)

Stoyan, Y.G., Yaskov, G., Scheithauer, G.: Packing of various radii solid spheres into a parallelepiped. Cent. Eur. J. Oper. Res. **11**(4), 389–407 (2003)

Stoyan, Y.G., Zlotnik, M.V., Chugay, A.M.: Solving an optimization packing problem of circles and non-convex polygons with rotations into a multiply connected region. J. Oper. Res. Soc. **63** (3), 379–391 (2012)

Sykora, A.M., Álvarez-Valdés, R., Tamarit, J.M.: Branch and cut algorithms to solve nesting problems. In: 8th ESICUP Meeting, Copenhagen, Denmark, 19–21 May 2011

Suhl, U.H.: Solving large-scale mixed integer programs with fixed charge variables. Math. Program. **32**(2), 165–182 (1984)

Sutou, A., Dai, Y.: Global optimization approach to unequal global optimization approach to unequal sphere packing problems in 3D. J. Optim. Theor. Appl. **114**(3), 671–694 (2002)

Teng, H., Sun, S., Liu, D., Li, Y.: Layout optimization for the objects located within a rotating vessel a three-dimensional packing problem with behavioural constraints. Comput. Oper. Res. **28**(6), 521–535 (2001)

Terashima-Marín, H., Ross, P., Farías-Zárate, C.J., López-Camacho, E., Valenzuela-Rendón, M.: Generalized hyper-heuristics for solving 2D regular and irregular packing problems. Ann. Oper. Res. **179**, 369–392 (2010)

The MathWorks: MATLAB. The MathWorks, Inc., Natick, MA. www.mathworks.com (2012). Accessed 30 Aug 2013

Torquato, S., Jiao, Y.: Dense polyhedral packings: Platonic and Archimedean solids. Phys. Rev. E **80**, 041104 (2009)

Van Roy, T.J., Wolsey, L.A.: Solving mixed integer programming problems using automatic reformulation. Oper. Res. **35**, 45–57 (1987)

Vielma, J.P., Nemhauser, G.L.: Modeling disjunctive constraints with a logarithmic number of binary variables and constraints. Math. Program. **128**(12), 49–72 (2009)

Wang, Z., Li, K.W., Levy, J.K.: A heuristic for the container loading problem: a tertiary-tree-based dynamic space decomposition approach. Eur. J. Oper. Res. **191**(1), 86–99 (2008)

Wang, P.C., Tsai, J.F.: A superior piecewise linearization approach for assortment problems. In: 24th European Conference on Operations Research, Lisbon, Portugal, 11–14 July 2010

Weismantel, R.: Hilbert bases and the facets of special knapsack polytopes. Math. Oper. Res. **21**, 886–904 (1996)

Weisstein, E.W.: Grid Graph. MathWorld—A Wolfram Web Resource. http://mathworld.wol fram.com/GridGraph.html (2012). Accessed 30 Aug 2013

Williams, H.P.: Model Building in Mathematical Programming. Wiley, London (1993)

Wolsey, L.A.: Strong formulations for mixed integer programming: a survey. Math. Program. **45**, 173–191 (1989)

Wolsey, L.A.: Valid inequalities for mixed integer programs with generalised upper bound constraints. Discrete Appl. Math. **25**, 251–261 (1990)

Wolsey, L.A.: Strong formulations for mixed integer programs: valid inequalities and extended formulations. Math. Program. **97**(1–2), 423–447 (2003)

Yaman, H.: Polyhedral analysis for the two item uncapacitated lot sizing problem with one way substitution. Discrete Appl. Math. **157**, 3133–3151 (2009)

Yeung, L.H., Tang, W.K.: A hybrid genetic approach for container loading in logistics industry. IEEE Trans. Ind. Electron. **52**(2), 617–627 (2005)

Zhang, H.: Packing: scheduling, embedding, and approximating metrics. In: Laganá, A., Gavrilova, M.L., Kumar, V., Mun, Y., Tan, C.J., Gervasi, O. (eds.) Computational Science and Its Applications ICCSA, pp. 764–775. Springer, Berlin (2004)

Zhang, D., Kang, Y., Deng, A.: A new heuristic recursive algorithm for the strip rectangular packing problem. Comput. Oper. Res. **33**, 2209–2217 (2006)

Index

G. Fasano, *Solving Non-standard Packing Problems by Global Optimization and Heuristics*, SpringerBriefs in Optimization, DOI 10.1007/978-3-319-05005-8, © Giorgio Fasano 2014